Audio amp projects

RA Penfold

PC Publishing
Export House
130 Vale Road
Kent TN9 1SP
UK

Tel 01732 770893
Fax 01732 770268
email pcp@cix.compulink.co.uk
website http://www.pc-pubs.demon.co.uk

First published 1997

ISBN 1 870775 52 X

British Library Cataloguing in Publication Data
A catalogue record for this book is available from the British Library

Printed in Great Britain by Bell and Bain, Glasgow

Preface

Audio projects have been a popular aspect of electronic gadget building since the days of valve circuits. In those days the emphasis was very much on amplifiers for the hi-fi enthusiast, but these days there is a wide range of audio related projects that can be built by the home constructor. These include such things as graphic equalisers, audio limiters, filters, and dynamic noise reduction units. While most of these devices are of interest to the hi-fi enthusiast, many also find use in other fields. For example, many audio projects are of great interest to the video enthusiast, including such things as audio mixers, graphic equalisers, and dynamic noise limiters. Equipment of this type can be constructed at lower cost than buying ready made alternatives, and you have the added fun of building the projects.

This book features a useful range of audio projects that are all easy to build, and some are so simple that even complete beginners at electronic project construction can tackle them with ease. The preamplifier circuits and the tone controls are well suited to beginners, as is the small power amplifier project. Chapter 1 is primarily intended for beginners, and covers component identification, soldering, basic construction techniques, and troubleshooting. Stripboard layouts and wiring diagrams are provided for every project, together with notes on any unusual aspects of construction, where appropriate. The mechanical side of construction has largely been left to individual constructors to sort out, simply because the vast majority of project builders prefer to do their own thing in this respect.

None of the designs require the use of any test equipment in order to get them set up properly. When any setting up is required, the procedures are very straightforward, and they are described in detail.

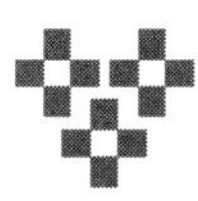

Contents

1 Getting started

The projects featured in this book should present few difficulties to anyone who has some previous experience of electronic project construction. In fact many of the projects are simple enough for complete beginners at electronics, but there is some essential background information that must be acquired before anyone new to this type of thing starts soldering in earnest.

In this chapter it is not possible to provide a complete course on electronic components, methods of construction, etc. What is provided is an introduction to the components, construction techniques, etc. that are needed in order to build the particular projects described in this book.

With this information, a certain amount of skill, some common sense, and a bit of ingenuity, practically anyone should be able to build the more simple of the projects. If you are not a very practical sort of person, then I would be misleading you to say that you could still complete these projects successfully. It would not be true either, to say that an average beginner could successfully tackle the more complex projects featured in this book. However, provided you are not completely useless when it comes to manual skills, and you start with the simple projects first, it should not be too difficult to progress successfully to the more complicated projects.

If you do not already have one of the larger electronic component mail order catalogues, then I would strongly advise getting at least one of these before starting to buy any of the components for a project. Here are just some of the component catalogues you will find well worth having:
MPS (Maplin)
Cirkit
Electromail
Electrovalue

Components

The range of electronic components currently available to amateur users is vast. There are literally thousands of different components available. Fortunately, only a fairly small percentage of these are used in the projects featured here. This makes it relatively easy to obtain and identify the components.

If you are lucky enough to live near a branch of one the large electronic component retailers, then you may be able to obtain everything you need locally. Even if this should be the case, it is still worthwhile having one or two large electronic component catalogues. These contain a mass of useful data, and also have lots of photographs or drawings of the components. These are very useful for beginners, as they make it relatively easy to find the right components. To put things another way, it reduces the risk of wasting time and money buying the wrong thing. If you do not live near to a suitable shop, then a big electronic components catalogue and mail order buying represent the only practical method of obtaining the components you will need for these projects.

Resistors

Resistors must be the most common of electronic components. Practically every electronic project uses some of these, and in most cases they represent about half the components in a project. There are numerous different types of resistor available, but for most of the projects in this book ordinary carbon film resistors will suffice. For the ultimate in performance in low noise preamplifiers it might be worthwhile investing in higher quality resistors such as the metal film variety, but this is not essential. The extra quality of metal film resistors is unlikely to bring any significant benefits with the other projects featured here, but it will obviously not do any harm either. Metal film resistors are perfectly suitable for these projects, but where carbon film types are available they will do the job just as well but at a much lower cost.

Resistance

The values of resistors are specified in ohms. The Greek letter omega is still used as an abbreviation for ohms, but these days a capital 'R' is more commonly used. Thus a 10 ohm resistor will often be referred to as a 10R component. In fact a value of 10R, particularly on circuit diagrams, is sometimes just given as '10'. An ohm is a very small unit, and many of the resistors used in electronics have values of thousands of ohms, or even millions of ohms. One thousand ohms equal one kilohm, and the abbreviation 'k' is often used for kilohms. A million ohms is a megohm, or just 'M'.

A 33000 ohm resistor would therefore normally have its value given as 33 kilohms, or just 33k. A 1000000 ohm resistor has a value of 1 megohm, or just 1M for short.

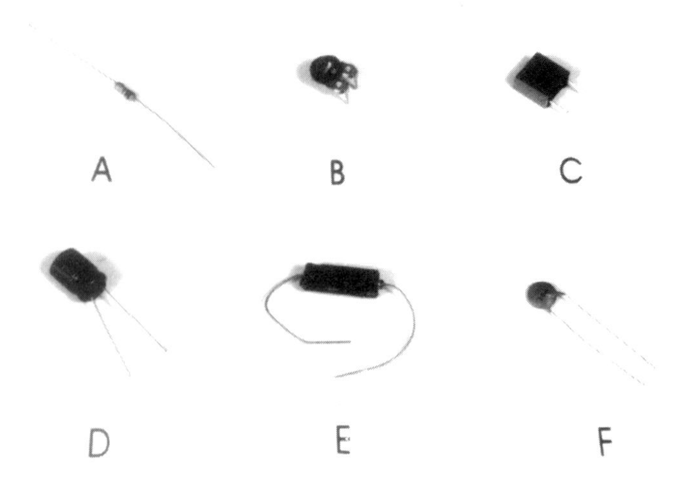

Photo 1.1
A Small resistor
B Preset resistor
C Printed circuit mounting capacitor
D Radial electrolytic capacitor
E Axial electrolytic capacitor
F Disc ceramic capacitor

Resistors are available only in certain values, known as preferred values. This is what is generally known as the E24 series of values:

1.0	1.1	1.2	1.3	1.5	1.6
1.8	2.0	2.2	2.4	2.7	3.0
3.3	3.6	3.9	4.3	4.7	5.1
5.6	6.2	6.8	7.5	8.2	9.1

This series of values might look a little odd, but it operates on the principle of having each value about 10% higher than the previous one. This ensures that whatever the calculated value for a resistor might be, there will always be an actual value available that is within a few percent of this value.

Note that components are not just available in these values, but also in their decades. In other words, as well as (say) 2.2 ohm resistors, there are also 22R, 220R, 2k2, 22k, 220k, and 2M2 types available.

Resistors are not generally available with values higher than 10M, and so you will probably not find 22M types listed in components catalogues.

TIP

The letter which indicates the unit in use is often used to indicate the decimal point as well. Thus a 4R7 resistor is a 4.7 ohm type, a 5k6 resistor is a 5.6 kilohm (5600 ohm) component, and a 1M8 resistor is a 1.8 megohm (1800000 ohm) type. The point of this system is that it enables values to be represented using as few digits as possible. This is especially useful when trying to find space for the labels on circuit diagrams.

Tolerance

Resistors have a tolerance rating, and this is usually 5% (possibly 10% on the higher values). The actual value of a resistor is never precisely its marked value. The tolerance rating is merely an indication of the maximum error. For example, a 100k 5% resistor would have an actual value of between 95k (100k minus 5%) and 105k (100k plus 5%). Ordinary 5% and 10% resistors are fine for the projects in this book. Metal film resistors, incidentally, mostly have a tolerance rating of just 1% or 2%. These are a bit over-specified for most purposes, but are otherwise perfectly suitable for use in these projects.

Power rating

Last, and by no means least, resistors have a power rating. This is basically just the maximum power in watts that the resistor can withstand. Practical power ratings tend to be a bit misleading as they are often quoted for different operating conditions. A 0.6 watt resistor might have a power rating that represents something close to the point where the component burns up, whereas a 0.25 watt type might have a rating that represents a very safe maximum dissipation figure. Their real power handling ability might therefore be quite similar, despite the fact that the power ratings are very different.

For these projects 0.25 watt resistors will suffice. Higher power types, such as 0.33, 0.4, 0.5, and 0.6 watt types are also suitable, provided they are miniature types. Resistors such as old style 0.5 and 1 watt resistors are fine from the electronic point of view, but you are unlikely to be able to fit them into the component layouts provided here!

Colour coding

It is very unusual for resistors to have their values marked in numbers and letters. With high power types you do sometimes find that the value is written on as 4R7, 68R, or whatever, plus a letter to show the tolerance. It is many years since I last saw any small resistors which use this method though. The standard method of value marking is to have four coloured bands. This method of coding works in the manner shown in Figure 1.1 and Table 1. In theory, the first band is the one nearest to one end of the resistor's body. I would have to say that with all the small resistors in my spares box the two end bands are an equal distance from their respective ends of the body. However, it is still easy enough to tell which band is which, as the fourth band is well separated from the other three.

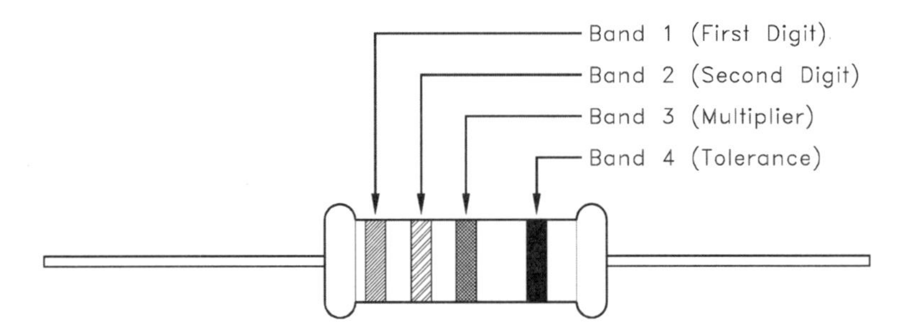

Figure 1.1 The standard four band method of resistor colour coding

Table 1 Resistor colour code

Colour	Band 1/2	Band 3	Band 4
Black	0	x1	-
Brown	1	x10	1%
Red	2	x100	2%
Orange	3	x1000	-
Yellow	4	x10000	-
Green	5	x100000	0.5%
Blue	6	x1000000	0.25%
Violet	7	-	0.1%
Grey	8	-	-
White	9	-	-
Gold	-	-	5%
Silver	-	-	10%
None	-	-	20%

The first two bands indicate the first two digits of the value. For example, if these bands are respectively green and blue, the first two digits are 5 and 6, as can be seen from the table. The third band is the multiplier, and this basically just indicates the number of zeros that must be added to the first two digits in order to give the full value. For instance, if the third band is orange, this indicates that the first two digits must be multiplied by 1000, or that three zeros must be added in other words. Thus in our example value, we have 56 plus three zeros, or a value of 56000 ohms (56k). Band number four indicates the tolerance of the component. This will usually be gold, which indicates a tolerance of five percent.

There is a slight problem with resistor colour coding in that there are also a couple of five band codes in use. These are closely based on the standard four band type. With one of these five band codes you have what is basically the ordinary four band code, plus an extra band which indicates the temperature coefficient. The latter is a measure of how much the component's resistance changes with variations in temperature. This is something that is not normally of any interest, and the extra band can be ignored. There is an alternative five band code which has the first three bands to indicate the first three digits of the value. The last two digits indicate the multiplier value and tolerance in the normal way. This method of coding seems to be quite common for metal film resistors which have a tolerance rating of 1%, but is little used apart from this. If a retailer is selling resistors with one of these five band codes, then their catalogue should give details of the code in use.

Obviously there is no difficulty in determining the value of a resistor which simply has the value written on its body. The only potential source of confusion is the letter which indicates the tolerance. This coding is very straightforward, and operates in the manner shown in this table.

INFO

This colour coding system can be a little confusing at first, but it has the advantage that the value markings are not easily obliterated. Even if the coloured bands should become chipped or worn, there will probably still be enough left to permit the value to be reliably deciphered. With tiny lettering on miniature resistors, it would be difficult to read the values even if the lettering was in perfect condition!

Tolerance codes for resistors

Code Letter	Tolerance
F	1%
G	2%
H	2.5%
J	5%
K	10%
M	20%

Potentiometers

A potentiometer is a form of resistor, but one where the value can be varied. A normal potentiometer has a control shaft which is fitted with a control knob. There is also a mounting nut and bush so that it can be fixed to the front panel of the case. For the projects featured here it is the small carbon potentiometers that are required, not high power types such as wire-wound potentiometers. Usually the mounting bush is for a 10 millimetre diameter hole, and the control shaft is 6 millimetres in diameter. However, some modern potentiometers have smaller mounting bushes and (or) shafts. Some, for instance, have the standard 6 millimetre shaft diameter, but require a mounting hole of just 7 millimetres in diameter. These miniature types are suitable for the projects featured here provided you remember to make the smaller mounting holes that they require. It is probably best to avoid types which have a control shaft diameter of other than 6 millimetres unless you are sure that you can obtain control knobs to fit them.

Log and lin potentiometers

As far as their electrical characteristics are concerned, there are two types of potentiometer. These are the logarithmic ('log') and linear ('lin') types. A linear potentiometer is the normal kind, where setting the control at a roughly mid-setting gives about half maximum resistance. A logarithmic potentiometer has a very non-linear resistance characteristic. Adjustment towards one end of the track has little effect, while adjustment at the opposite end results in large changes in value. Logarithmic potentiometers are mainly used as volume controls, while linear potentiometers are used for practically all other potentiometer applications.

The components lists always specify which type is needed. Using the wrong type will not prevent a project from working, but you will get some strange control characteristics. This can make the projects difficult to use properly, so I would strongly recommend that the specified types should always be used.

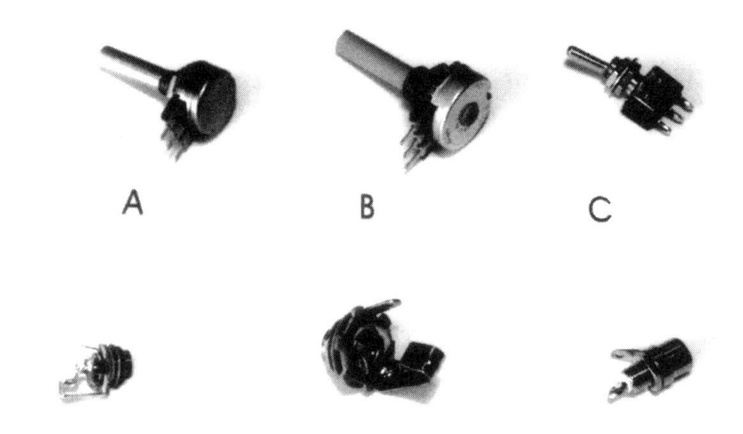

Photo 1.2
A Small potentiometer
B Standard potentiometer
C SPDT miniature toggle switch
D 3.5 mm jack socket
E Standard jack socket
F Phono socket

Preset potentiometers

Preset potentiometers are physically very different from the standard variety. These are small components, and are often of open construction. Fortunately, it seems to be increasingly common for preset potentiometers to have a plastic outer casing to keep dust and other contamination away from their inner workings. Only one project featured here requires a preset potentiometer, and this is the dynamic noise limiter. This requires a sub-miniature (0.1 or 0.15 watt) horizontal mounting preset. Other types are suitable electrically, but will probably not fit into the component layouts properly, and could be very expensive in the case of high quality multi-turn components.

Preset potentiometers are generally available only as linear types incidentally. Because of this, components lists do not normally specify whether a preset should be a logarithmic or a linear type, and component catalogues do not normally specify the type either.

Note that colour coding is not normally used for marking potentiometer values. Instead, the value is simply marked as '4k7 lin', or whatever. In some cases a letter 'A' after the value is used to denote a logarithmic component, or a letter 'B' is used to shown that it is a linear type. Colour coding is occasionally used for marking values on preset resistors. This coding is normally in the form of three coloured dots which indicate the value in the same way as the first three bands of a normal resistor colour code.

TIP

Unless stated otherwise, preset potentiometers can be assumed to be linear types.

Capacitors

Capacitors represent another type of component that is used in large numbers in electronic circuits. In a few projects they actually outnumber the resistors. The values of capacitors are in farads, but one farad is a massive amount of capacitance. Therefore, most 'real world' capacitors have their values marked in microfarads, nanofarads, or picofarads. The table shows the relationship between these three units of measurement. A microfarad is one millionth of a farad incidentally. The abbreviations 'u', 'n', and 'p' are often used for microfarads, nanofarads, and picofarads respectively.

Capacitor types

There are numerous different types of capacitor available. Here we will only consider the types that are relevant to the projects in this book. Where low values are called for, by which I mean values of under one nanofarad, polystyrene capacitors are suitable. These are normally axial components, which look a bit like fat resistors. The values are usually marked on the body in the same fashion that the value is marked on the

Capacitance

Unit	*Microfarads*	*Nanofarads*	*Picofarads*
Microfarads	1	1000	1000000
Nanofarads	0.001	1	1000
Picofarads	0.000001	0.001	1

circuit diagram or in the components list (e.g. '470p', '220p', '47p', etc.). Some polystyrene capacitors have one end of their body coloured red, but strictly speaking these are not polarised components, and they can be connected either way round. In radio frequency applications it can be advantageous to connect a polyester capacitor a certain way round, but this does not apply to the audio frequency circuits featured here.

Ceramic plate capacitors are also suitable, but in general are not my first choice for use in audio projects. Some of these capacitors have quite high tolerance ratings (20% or more): capacitors having such high tolerances are not suitable for use in these projects. Ceramic plate capacitors which have values of more than 100 picofarads often have slightly cryptic value markings, with the value being given in nanofarads. Thus 'n10' is 0.10 nanofarads or 100 picofarads, and 'n39' is 0.39 nanofarads or 390 picofarads. This is basically the usual method of using the units indicator to also show the position of the decimal point, but no leading zero is included.

There can be a problem with ceramic plate capacitors in that some of the components currently being sold have very short leadout wires indeed. In some cases these will fit into the component layouts without any difficulty, but in a few cases it might be necessary to solder short extension wires to their leadout wires in order to fit them into the component layouts properly.

Polystyrene components are also suitable where values of under one nanofarad are called for. The only proviso here is that they must be physically small types. There should be no problem if modern components are used, and you do not order a type intended for operation at very high voltages. The values of small polystyrene capacitors are normally marked on the components in picofarads, possibly with the tolerance shown as well, or a code letter to show the tolerance. Ordinary 5% tolerance polystyrene capacitors are suitable for these projects.

For values from one nanofarad to around one microfarad the component layouts are mainly designed to take printed circuit mounting polyester capacitors. These have very short leadout wires that are really more like pins than normal leadout wires. This means that the component layouts have to be designed to suit components of a particular lead spacing. In this case a lead spacing of 7.5 millimetres (0.3 inches) is used for virtually all the medium value capacitors. Check the component layouts and be careful to order components that have the appropriate lead spacing.

Some component retailers now offer 'cased' versions of polyester capacitors instead of open construction types. These have tough plastic encapsulations, and there should be little risk in manipulating them into position.

It is only fair to warn you that the likely result of trying to manipulate polyester capacitors which have the wrong lead spacing into these component layouts is that leadout wires will become detached from the capacitors. In days gone by these polyester capacitors were very easily damaged in this way. Modern types seem to be much tougher, but they are still likely to be damaged if you bend the wires to suit a different lead spacing.

The values of polyester capacitors are normally marked on the components in nanofarads or picofarads, as appropriate. A 2.2 nanofarad component would therefore be marked '2n2', a 10n component would be marked '10n', and a 470nf (0.47uf) component would be marked 'u47' or '0u47'.

Mylar capacitors offer an alternative to polyester capacitors for values of up to about 47n. This is a form of printed circuit mounting capacitor, but Mylar capacitors normally have long leadout wires. This makes it easy to fit them into practically any component layout. Unfortunately, values of more than about 47n tend to be physically large, making them unsuitable for many of the board layouts featured in this book.

In a few cases ceramic capacitors are specified for values of around 100 nanofarads. These higher value ceramic capacitors have very inaccurate values, but they work well at high frequencies. This makes them well suited to certain applications, including decoupling types. It does not matter which type of ceramic capacitor is used, but disc types are the probably most widely available, and are generally the cheapest. Obviously you should avoid very high voltage types, or any special ceramic types which are expensive or physically quite large. Do not use ceramic capacitors where the components list calls for polyester capacitors.

The value of ceramic capacitors is often just marked in nano or microfarads. Probably the most common method of value marking though, is one which uses three numbers. The first two numbers are the first two digits of the value. The third number is the number of zeros that must be added to these in order to give the value in picofarads. For a 100 nanofarad capacitor the marking would be '104'. The first two digits of the value are '1' and '0', and four zeros must be added to these. This gives a value of 100000 picofarads, which is the same as 100 nanofarads.

Electrolytics

Ordinary capacitors are not a very practical proposition where high values of more than about one microfarad are required. They are expensive to produce, and tend to be physically quite large. For high values it is normal to use electrolytic capacitors. These are not without their drawbacks, such as relatively high tolerances and high leakage currents, but they are adequate for many purposes where high values are needed.

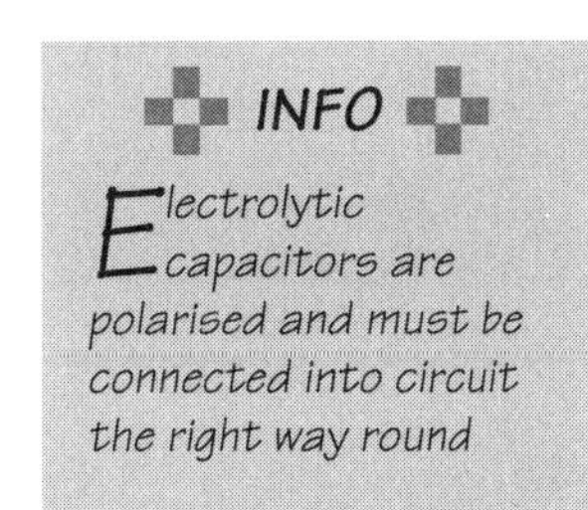

The most important difference between electrolytic and non-electrolytic capacitors is that the electrolytic type are polarised. In other words they have positive and negative terminals, and must be connected into circuit the right way round if they are to function properly. Ordinary capacitors such as polyester and ceramic types, and resistors, can be fitted either way round.

Identifying the positive and negative leadout wires should not be difficult, since there are usually '+' and '-' signs marked on the body of an electrolytic capacitor which clearly show which lead is which. There seems to be a tendency these days towards only marking one lead or the other, but this is obviously all you need in order to get the component fitted round the right way.

Physically there are two different types of electrolytic capacitor. These are the radial and axial types. These are actually general terms that are applied to other types of component, but which are mainly encountered when dealing with electrolytic capacitors. An axial type is the usual tubular bodied component having a leadout wire protruding from each end. These would normally be mounted horizontally on the circuit board. A radial component has both leadout wires coming from the same end of the component, and it is intended for vertical mounting. It is usually possible to fit a radial capacitor into a layout that is designed for an axial type, or vice versa. Some careful forming of the leadout wires may be needed, or some extension wires might have to be added to the leadout wires.

It is best to avoid this type of thing though. Apart from the fact that it will give some slightly scrappy looking results, it is likely to result in components being something less than firmly fixed in place. This leaves the board vulnerable to problems with broken leads or short circuits from one component lead to another.

You will notice that axial electrolytics have an indentation around one end of the component's body. This is used to indicate the end of the component from which the positive (+) lead emanates. The + and (or) – markings are usually included as well, but the indentation enables you to see the polarity at a glance, making these markings largely superfluous.

It is perhaps worth pointing out that radial capacitors are also know as PC (printed circuit) or PCM (printed circuit mounting) capacitors. This is actually a term which can be applied to any component which is intended for vertical mounting on a printed circuit board. In the early days of electronics all two lead components were of the axial type. When printed circuits became along, these vertically mounting components were designed specifically to give compact component layouts with the new method of construction. Hence these became known as printed circuit mounting components, and eventually just PC or PCM components.

Voltage ratings

With non-electrolytic capacitors you do not normally need to worry about the maximum voltages they can safely handle. The voltage ratings are generally around the 100 volt mark, or in some cases even higher. The projects in this book mostly operate on a nine or 12 volt battery supply.

The situation is slightly different with electrolytic capacitors. The smaller values generally have voltage ratings of about 50 volts or more, but the higher values can have voltage ratings as low as 6 volts, or possibly even 3 volts. The components lists specify the sort of voltage rating that you are likely to encounter in components catalogues. This means that low value electrolytic capacitors are given a voltage rating of 50 volts, even though their maximum operating voltage might only be a few volts. There is little point in specifying a capacitor as a 2u2 5 volt type when component catalogues only have components of this value with voltage ratings of around 50 to 100 volts.

Higher value electrolytic capacitors are available with relatively low voltage ratings. Consequently, the voltage ratings for these are generally much closer to their actual operating voltage. It is not essential to use electrolytic capacitors having exactly the same voltage ratings that appear in the components lists, but unless you are sure you know what you are doing, do not use components having lower voltage ratings. For example, it is in order to use a 1u 63V or 1u 100V component in place of a 1u 50V type, but a 100u 10V capacitor should not be used where a 100u 16V type is specified. The only proviso is that the component must be small enough to fit into the component layout correctly. For instance, a 220u 100V capacitor will work in place of a 220u 16V type, but could be two or three times bigger on each dimension.

Diodes

A diode is a type of semiconductor (like transistors and integrated circuits), but is the most simple type of semiconductor. It acts like a sort of electronic valve which enables an electric current to flow in one direction, but not the other. These components are also called rectifiers. Diodes are used in a few of the projects featured in this book, and rectifiers are used in the power supply unit for the 20/32 watt amplifier.

Obviously a diode or rectifier must be connected the right way round if it is to let the current flow in the right direction. The standard method of polarity marking for diodes and the smaller rectifiers is to have a band marked around one end of the component's body. This indicates the cathode (+) leadout wire. The component layout diagrams in this book show this band on all diodes, so all you have to do is fit these components onto the board so that they match up with the diagrams in this respect. You do not really need to worry about what the terminals are called. The rectifiers used in the power supply project are a large type, and they have the cathode end of the body slightly pointed rather than having a band marked around the body (although the band might be present as well). The relevant wiring diagram clearly shows this tapering of the rectifiers, making it obvious which way round each one is connected.

The only difference between a rectifier and a diode is that the former is used in medium and high power applications, whereas a diode is used for small signals.

Type numbers

Diodes, like all semiconductors, do not have values. Instead they have type numbers, and the data sheet for each component shows its electrical ratings and characteristics. To make it easier to find the particular semiconductor you require, most component catalogues list diodes, rectifiers, transistors, etc. separately.

The only diodes used in these projects are the 1N4148 and OA91. These are both very common types that you should be able to find listed in any electronic components catalogue. The equally common 1N914 is a suitable substitute for the 1N4148. Similarly, rectifiers also have type numbers, and the power supply uses 1N5402 rectifiers. The 1N5402 is listed in the rectifier section of most electronic component catalogues, but any rectifier have a peak inverse voltage (p.i.v.) rating of 200 volts or more, and a current rating of three amps or more should work well in the power supply circuit.

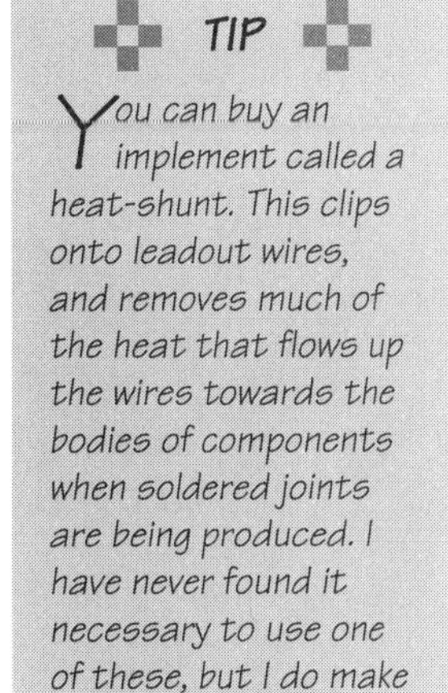

TIP

You can buy an implement called a heat-shunt. This clips onto leadout wires, and removes much of the heat that flows up the wires towards the bodies of components when soldered joints are being produced. I have never found it necessary to use one of these, but I do make sure that every soldered joint is completed fairly rapidly.

Heat damage

It is worth mentioning here that semiconductors are rather more vulnerable to heat damage than are most other electronic components. Modern silicon devices are somewhat more hardy in this respect than the old germanium based devices. However, even the silicon based devices need to be treated with due respect when they are being soldered into circuit.

The only germanium semiconductors used in these projects are the OA91 diodes used in the dynamic noise limiter, and extra care should obviously be taken when soldering these in place.

If it becomes evident that a joint is not proceeding well, it is better to abandon it and try again once everything has cooled down, rather than pressing on and possibly damaging the component.

It is worth bearing in mind that no electronic components are totally heat-proof. Even with simple components such as resistors you can cause damage if you keep the bit of the soldering iron in place for so long that the component becomes discoloured and smoke begins to rise! Components will sometimes survive this sort of treatment, but often with shifted values or other problems that will degrade performance. Try to complete all soldered joints reasonably quickly, preferably taking no more than a second or two.

Transistors

Normally in a book of this type I would include a section dealing with transistors. In this case there is no point, since there are no discrete transistors in any of the designs! There are plenty of transistors hidden away inside the integrated circuits, but no individual transistors are used in any of the designs.

Integrated circuits

Modern electronic circuits tend to be extremely complex – even the simple ones! This might seem to be an impossible state of affairs, but it is made possible by integrated circuits. Strictly speaking each integrated circuit is a single component. However, an integrated circuit actually contains the equivalent of what could be anything from two to over one million components.

Although a project might only use a dozen components, two of these could be integrated circuits, with each one containing the equivalent of hundreds or thousands of components. This brings tremendous benefits to the electronics hobbyist. Many projects that would otherwise be impracticable are brought within the scope of the average constructor. Integrated circuits can be used to reduce component counts to a realistic level, and also help to keep down costs. Very simple projects that actually do something useful become a practical proposition.

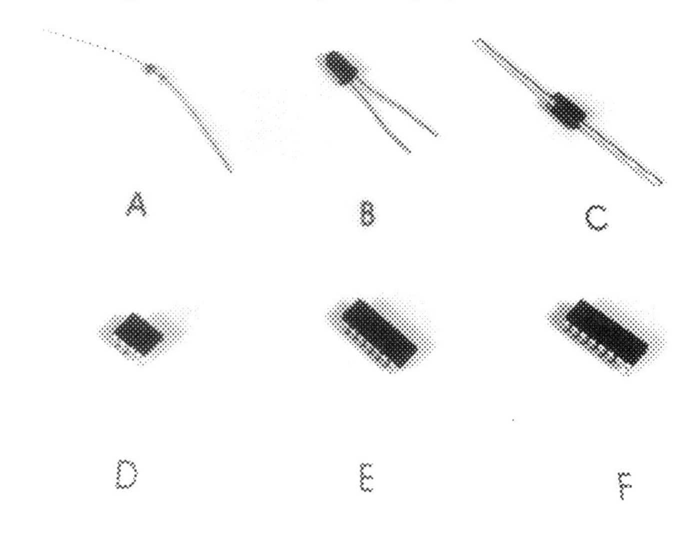

Photo 1.3
A Diode
B LED (5 mm dia)
C Large rectifier
D 8-pin d.i.l. IC
E 14-pin d.i.l. IC
F 16-pin d.i.l. IC

Some integrated circuits are quite cheap, and cost a matter of pence rather than pounds. Where possible, the projects in this book are based on low cost devices that are available from practically every electronic component supplier. In a few cases there is no choice but to opt for high quality devices that are less widely available and more expensive. Most of the circuits in this book are based on integrated circuits called 'operational amplifiers'. Where very low noise and distortion operational amplifiers are specified it is possible to use 'bog standard' devices instead. However, the saving in cost may not be that great, and the reduction in performance could be very substantial. In the more demanding applications it is definitely a good idea to use superior grade operational amplifiers that will provide really good results.

Linear and non-linear ICs

There are many hundreds of integrated circuits listed in most components catalogues. In order to make it easier to find the devices you require it is standard practice for the integrated circuits to be listed in several categories. Most of the devices used in these projects are 'linear' integrated circuits. The only non-linear type used in these projects is the 4007UBE, which is a CMOS logic device. Practically any component catalogue should have a CMOS logic section which lists this device. As there are many instances of totally different integrated circuits which very similar type numbers, you need to be careful when ordering these components.

Another point to bear in mind is that the same device might be produced by two or more manufacturers, but under slightly different type numbers. There is a popular linear device called the uA741C. I believe that this was the type number used by the original manufacturer, and it is still specified as such in many components lists. However, it is now produced by several semiconductor manufacturers, under type numbers such as LM741C, MC741C, and CA741C.

In components catalogues a device such as this might be listed under one specific type number, or even two or three different type numbers if the retailer stocks components from more than one source. In many cases though, popular integrated circuits which are produced by several manufacturers are simply listed under a sort of generic type number, which in this example would just be '741' or '741C'. As far as I am aware, the integrated circuits used in this book do not have generic type numbers, and should be available under the specific type numbers provided in the components lists. The actual devices supplied might, in some cases, be marked with a slightly different type number. This is usually where the component has been obtained from a so-called 'second source' manufacturer. This is where the original manufacturer has licensed one or more companies to produce the component. Second source suppliers sometimes use the original type number, but it is not uncommon for them to alter it slightly to fit in with their own range of type numbers. Really, the component supplier should make it clear if a substitute has been supplied. If in doubt, it is probably best to query the incorrect type number with your component supplier.

Encapsulation

Integrated circuits come in a variety of sizes and shapes. Many devices are available in more than one encapsulation. However, in component catalogues you will normally only find the d.i.l. (dual in-line) versions listed. These are basically rectangles of black plastic which contain the silicon chip, with a row of metal pins running along each side of the plastic case. It is from these two lines of pins that the d.i.l. name is derived. Most integrated circuits have 8, 14 or 16 pins, but they can have anything from four to 40 or more pins. The projects featured here only have 8, 14 or 16 pins though.

Static electricity

Some integrated circuits are built using some form of MOS (metal oxide silicon) technology, and the practical importance of this is that they are sensitive to static voltages. A large static charge could probably damage any integrated circuit, but charges of this magnitude are not to be found in normal environments. MOS components can be damaged by relatively small static charges, such as those that tend to be generated by nylon carpets, clothes made from synthetic fabrics, etc. The risk of components being damaged in this way has perhaps tended to be exaggerated slightly. Manufacturers' warnings can give the impression that MOS devices will be instantly destroyed unless they are stored and handled under carefully controlled conditions using thousands of pounds worth of specialised equipment!

In reality the risk is normally quite small. Many electronics hobbyists do not bother with any special handling precautions at all when dealing with MOS components, and in the main get away with it. However, when dealing with the more expensive MOS components it would seem to be prudent to exercise reasonable care. In fact it is not a bad idea to take a few simple precautions even when dealing with the cheaper components.

The most important of these is to leave static-sensitive components in their anti-static packaging until it is time to fit them onto the circuit board. Virtually every MOS device is supplied in some form of anti-static packing. This is usually conductive foam, ordinary plastic foam with a metal foil covering, a plastic tube, or a blister pack with a metal covering on the backing card. The basic idea is to either insulate the component from static charges, or (more usually) to short circuit all the pins together. This second method does not keep static charges away from the components, but it does ensure that dangerous voltages can not exist from one pin to another. It is a large voltage difference across the pins of a device that can actually cause damage.

Another important precaution is to avoid soldering MOS integrated circuits direct to the component panel. It is much safer if a suitable d.i.l. integrated circuit holder is soldered to the board, and the MOS Integrated circuit is then plugged into the holder. It is best to leave out the integrated circuits until a project has been completed in all other respects. This ensures that the components are kept away from the heat of the iron, or any electric currents that may 'leak' from the soldering iron.

Some MOS devices have built-in protection circuits that render handling precautions unnecessary, and these may not be supplied in anti-static packing.

I would strongly urge the use of holders for all integrated circuits, regardless of whether or not they are MOS devices. If you should accidentally fit an integrated circuit the wrong way round, there is little difficulty in gently prising it free from the holder using a small screwdriver, and then fitting it the right way round. Desoldering a multi-pin component from a circuit board can be quite difficult even if you do have access to proper desoldering equipment. Presumably many of the constructors tackling these projects will not have access to anything more than some very basic desoldering gear.

The other main anti-static precaution is to simply avoid getting the devices in contact with any likely sources of high static voltages. For example, when constructing projects avoid wearing clothing made synthetic fibres. When fitting MOS devices into their holders, try to avoid touching the pins as far as possible. In practice it might be impossible to avoid this completely. You often have to pinch the two rows of pins inwards slightly in order to get the integrated circuits into their holders.

To make it easy to get integrated circuits the right way round they have an indentation next to pin 1, plus a 'U' shaped notch at this end of the component. Figure 1.2 shows the notch and the 'dimple', and as is the convention, the integrated circuit is shown viewed from above. Note that this is the opposite to transistors, which are normally shown as base views in leadout diagrams. It is a 14 pin d.i.l. device that is shown in Figure 1.2, but the same method is used for all normal d.i.l. integrated circuits.

When building projects you do not normally need to worry about the pin numbering. Just get the devices orientated so that the notches and 'dimples' match up with the markings in the component layout diagrams. It is only fair to point out that many integrated circuits seem to have just the notch or only the 'dimple', but not both. This does not really matter, since either one of them will enable you to get the device orientated correctly. Slightly confusingly, some devices have a sort of notch at the opposite end of the case to the proper one. This is presumably just some sort of moulding mark, and it is not usually difficult to see which is the proper notch and which is not. The phantom notch is usually much larger and less deep than the real one.

TIP

It is essential that integrated circuits are fitted the right way round. Get one of these components the wrong way round, and it could easily be destroyed when the project is switched on. Unfortunately, it is probably true to say that, in general, the more expensive the integrated circuit, the more vulnerable it is to damage from this sort of treatment.

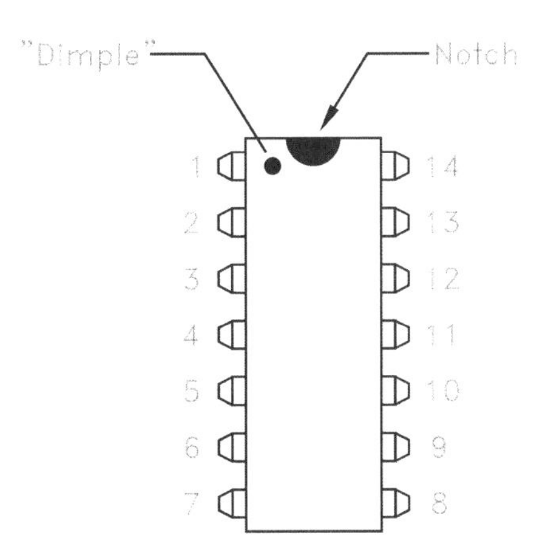

Figure 1.2 The notch and (or) dimple make it easy to fit integrated circuits the right way round

Sockets

Only three types of socket are used in these projects. These are standard 0.25 inch (6.35 millimetre) jack sockets, 3.5 millimetre jack sockets, and phono sockets. The type specified for each project is the one that I consider to be most appropriate, but the best type depends on the circumstances in which the equipment will be used. These projects should work using any type of audio connector, and the specified connectors can be changed to any type that better suits your particular set-up.

Standard jack sockets are available in two basic types, which are the open construction and insulated varieties. I used the open type for the projects in this book, but the insulated type is equally suitable. The open type requires a 9.5 millimetre diameter mounting hole. Note that the tag which connects to the 0 volt rail of the circuit connects internally to the metal mounting bush. Consequently, if the socket is mounted on a metal case or panel, it will connect the 0 volt rail of the circuit to the case or panel via the socket. This is often an asset, since having the case 'earthed' to the 0 volt rail helps to provide effective screening of the circuit. In certain circumstances it could give problems, but should not do so with the projects featured in this book.

Insulated sockets are an enclosed type having a plastic casing. This type of socket does not provide an earth connection to the case. This type of jack socket requires an 11 millimetre diameter mounting hole. Insulated jack sockets often have some switch contacts that are activated by inserting and removing the jack plug. These contacts are of no use in the current context and simply confuse matters.

These days many types of socket are available in chassis and printed circuit mounting varieties. It is the chassis mounting type that are required for these projects. In other words, sockets that are mounted on the case and wired to the circuit board using insulated leads. Printed circuit mounting sockets are intended to be fitted direct onto the circuit board, but the circuit boards featured here are not designed to accept this type of socket (which do not seem to have properly standardised pin arrangements anyway). Unless you know what you are doing, I would recommend the use of the open type, chassis mounting sockets, which have no switch contacts.

3.5 millimetre jack sockets are also available in chassis and printed circuit mounting varieties, and as open or insulated sockets. Chassis mounting sockets of open construction are the most readily available type, and it is these that are used in my prototypes. Other chassis mounting 3.5 millimetre jack sockets should be suitable provided you can work out the correct method of connection. Open 3.5 millimetre jack sockets are a bit confusing because they have three tags rather than two. The extra tag forms part of a switch that was originally intended to disconnect the internal loudspeaker when an earphone was plugged into the socket. The built-in switch serves no useful function with the projects featured here, and the extra tag is simply ignored.

Phono sockets are available in a variety of shapes and sizes. In this case the easiest type to use are the single chassis mounting sockets that require one fixing hole. All the sockets of this kind that I have used require a 6.35 millimetre diameter fixing hole. Gold-plated versions are available, but the standard (silver coloured) sockets are perfectly adequate for most purposes.

Switches

Most of the projects use one or more switches, and these are mostly very simple switches. In the components lists I have mainly specified sub-miniature toggle switches. A toggle switch is simply one that is controlled via a small lever (known as a 'dolly'). The sub-miniature versions of these switches are extremely small, reasonably cheap, and seem to be quite reliable these days. However, other types of switch, such as slider, rocker, and rotary types are also suitable. Bear in mind though, that with these other types of switch the mounting arrangements can be quite awkward. With rocker types in particular, you often have to make a rectangular mounting hole very accurately indeed. Make the cut-out fractionally too small and the switch will not fit into it, or marginally too large and the switch will not snap into place properly. With the slightest provocation the switch simply drops out the mounting hole. Sub-miniature toggle switches mostly require a single 6.3 millimetre diameter mounting hole, but the smallest types require 5.2 millimetre diameter mounting holes.

Another point to keep in mind is that other types of switch tend to be much larger than sub-miniature toggle types. This can make it difficult to find space for them inside the case, particularly in the case of a small project such as a simple preamplifier.

When dealing with switches you will encounter the terms s.p.s.t., s.p.d.t., d.p.s.t., and d.p.d.t. The first of these stands for single pole single throw, and this is the most simple type of switch. It has just two tags, and it is just a simple on/off type switch. A d.p.s.t. switch is a double pole single throw switch. This is basically just two s.p.s.t. switches controlled by a single lever, slider, or whatever. A switch of this type therefore has four tags.

INFO

s.p.s.t. *single pole single throw*
s.p.d.t. *single pole double throw*
d.p.s.t. *double pole single throw*
d.p.d.t. *double pole double throw*

An s.p.d.t. switch is a single pole double throw type. This type of switch has three tags. The centre tag connects to one of the other two, depending on the setting of the lever. These are also known as changeover switches, which is a fair description of their function. A d.p.d.t. switch is a double pole double throw switch, and this is two s.p.d.t. switches controlled in unison. This type of switch therefore has six tags. The components lists make it quite clear which type of switch is required.

The mains power supply unit for the 20/32 watt amplifier requires a d.p.s.t. switch to provide the on/off switching. With all the other switches the voltage rating is not of great significance, because they are operating at such low voltages. The on/off switch for this power supply operates at the 230 volt mains voltage, and it is important that its voltage rating is adequate for such a high operating voltage. Many miniature slider switches, toggle switches, etc. are only rated to operate up to about 100 volts, and are unsuitable. I prefer to use rotary switches for on/off switches in mains equipment, as it is easy to mount them reliably (they need a single 9.5 millimetre diameter fixing hole), and in my experience they are trouble-free.

Batteries

Virtually all the projects featured in this book are powered from a 9 or 12 volt battery. This helps to keep things simple, and for beginners it keeps things safe. The shock from 9 or 12 volt battery is so slight that you are totally unaware of it. The shock from the 230 volt mains supply can be lethal.

For some of the projects a small (PP3 size) battery is adequate, but with some a higher supply potential of 12 volts is preferable. Others require a higher capacity battery. Where higher capacity is required I recommend using six or eight HP7 size cells in one of the plastic battery holders that are available from the larger component retailers. With a nominal 1.5 volts per cell, six and eight cells respectively provide 9 and 12 volts. The connection to most of these holders is made via an ordinary PP3 type battery clip, but some holders have tags to take soldered connections.

As most readers will be aware, the battery must be connected the right way round. In the early days of semiconductors the likely result of getting the battery connected the wrong way round, even briefly, was the destruction of all the semiconductors in the circuit. Modern semiconductors are mostly more tolerant of the wrong supply polarity, but some integrated circuits will permit very high supply currents to flow if the supply polarity is incorrect. Even if a device should withstand these high currents, it will quickly overheat and be destroyed. Therefore, be very careful to get the battery clip connected with the right polarity.

In the wiring diagrams + and - signs are used to indicate the battery polarity. The red battery clip lead is the positive (+) one, and the black lead is the negative (-) one.

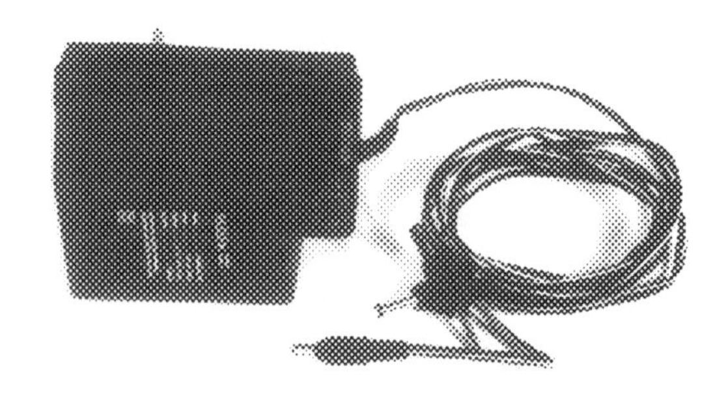

Photo 1.4 A regulated battery eliminator provides an economic alternative to batteries for most of the projects.

Stripboard

These projects are all based on stripboard, which is a form of proprietary printed circuit board. In some catalogues you may find it listed under the proprietary name 'Veroboard'. It consists basically of a piece of thin board which is brown in colour and made from an insulating material. It is drilled with holes on a 0.1 inch (2.54 millimetre) matrix, and strips of copper join up rows of holes. Figure 1.3 shows this general scheme of things. Something which tends to confuse some beginners is that the copper side of the board is often referred to as the underside, and the non-copper side is called the top side. It might seem reasonable to assume that the plain side would be called the underside. However, the components are fitted on the plain side of the board, and the board is normally mounted with this side uppermost.

Stripboard is sold in various standard sizes, but these projects do not use the board in these sizes. Therefore, boards of the correct size must be cut from larger pieces using a hacksaw. When cutting the board have the copper side facing upwards. Otherwise the copper strips that are cut might tend to rip away from the board. Cut along rows of holes rather

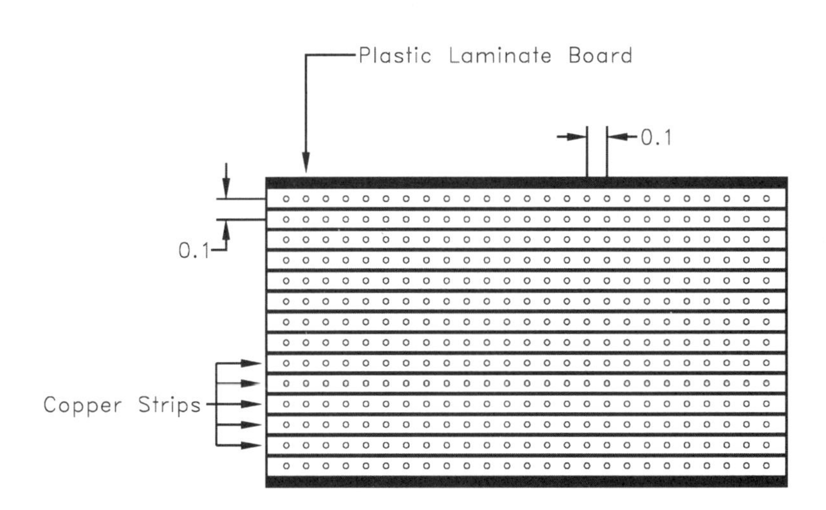

Figure 1.3 (a) Stripboard has narrow copper strips and 1mm diameter holes in a 2.54 mm matrix

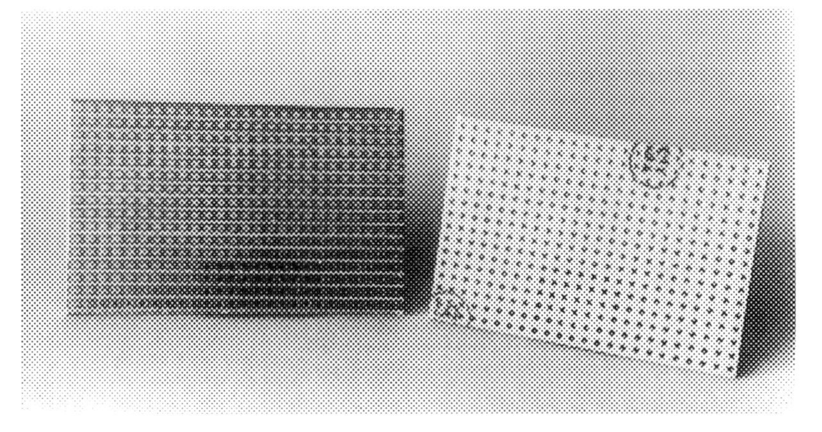

Photo 1.5 Copper strip side (left) and plain side (right) of stripboard

than trying to cut between rows (which is virtually impossible as they are so close together). This leaves rather rough edges, but these can easily be smoothed to a neat finish using a file. Some stripboards seem to be made of a rather brittle material that can easily crack and break when it is being trimmed to size. Always proceed carefully, using minimal force when cutting stripboard.

Figure 1.3 (b) Cuts in copper are shown as black circles like this

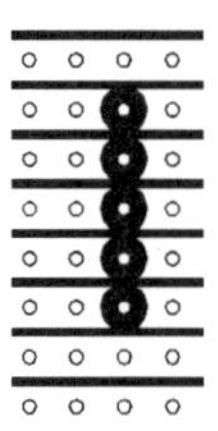

Most stripboard layouts require some cuts in the copper strips, so that some of the strips can be used to carry more than one set of connections. A special strip cutting tool is available, but the strips can be severed quite effectively using a twist-drill bit of about five millimetres in diameter. This can be used hand-held, since very little force is needed to cut through the copper strips. It is essential to cut deep enough to fully cut each strip across its full width, but do not get carried away and cut deeply into the board. Boards that have large numbers of cuts will be seriously weakened if you cut too deeply.

In most cases the relative orientations of the boards are obvious when you look at the component layout diagram and the underside view. However, to make quite sure that there is no confusion a triangle is marked against the same corner of the board in both the component and copper side views (Figure 1.4).

Make quite sure that you have the board the right way round before you start fitting any components. Removing components and refitting them can be difficult, and could damage the board.

Figure 1.4 The triangle is placed against the same corner of the board in the top and underside views

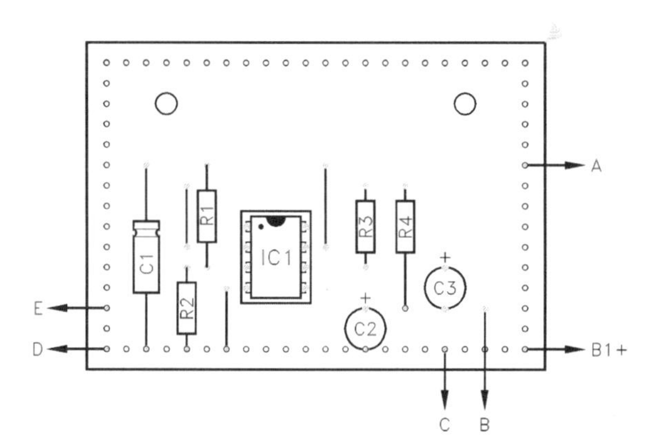

Soldering

The component leads are threaded through the appropriate holes on the non-copper side of the board, and the components are pressed hard against the board. The leads are trimmed on the underside of the board using wire clippers, and then soldered to the copper strips. The leads should be trimmed so that about two or three millimetres of wire protrudes on the underside of the board. The solder should then flow nicely over the end of the leadout wire and copper strip to produce a sort of mountain shaped joint, as in Figure 1.5(a).

If you end up with a blob shaped joint, as in Figure 1.5(b), then it is likely that you have a 'dry' joint which is not making a proper electrical connection between the lead and the copper strip. Other tell-tale signs of a bad joint are lots of burnt flux around the joint, and a dull, crazed appearance to the solder instead of the normal shiny finish. If you end up with a suspect joint, it is best to remove the solder, clean up the end of the leadout wire and the copper strip around it using a small file, and then try again.

The most important point to keep in mind when soldering components to a circuit board is that the bit of the iron should be applied to the joint first, and then some solder should be fed in. This gets the joint hot before

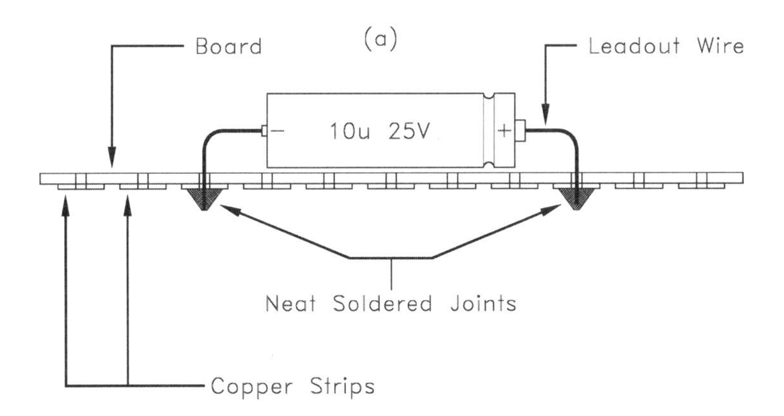

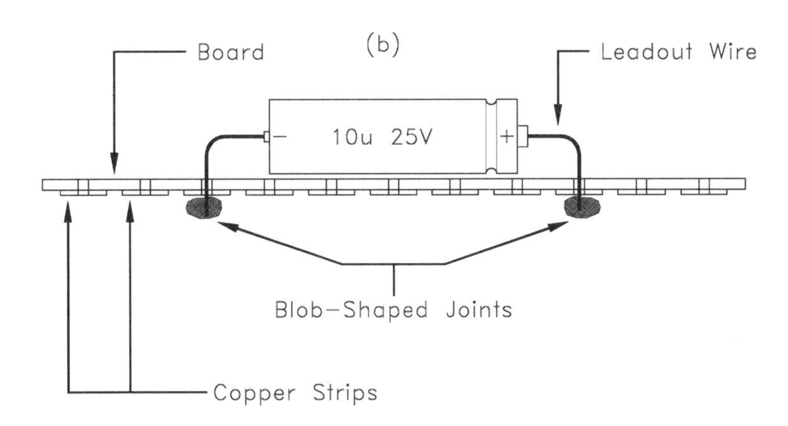

Figure 1.5 (a) A component connected with good soldered joints, and (b) joint of dubious quality

I would strongly urge newcomers to electronic project construction to practice soldering on a small piece of stripboard using some pieces of wire and resistors, prior to trying to build their first project. This may seem a bit wasteful, but the cost of the materials used in this exercise will probably be no more than about a pound. It could easily prevent you from wasting several pounds worth of components and a lot of time on a first project which becomes a total failure.

the solder is applied, which helps the solder to flow over the joint properly. What is definitely the wrong way of tackling this type of joint is to put some solder onto the iron and to then try to apply it to the lead and copper strip. The solder contains cores of flux which help the solder to flow properly over the joint. This flux tends to rapidly burn away if solder is placed on the bit of the iron.

When you try to transfer the solder to the joint there are then two problems. Firstly, with little flux in the solder it will not flow readily over the leadout wire and copper strip. Secondly, the joint has not been preheated, which also tends to restrict the flow of the solder. The result is usually a reluctance for the solder to leave the iron, and if it does, it usually just produces a blob on the leadout wire. This gives a joint which is highly unreliable both electrically and physically. Although this may seem to be an easy way of soldering, it is very ineffective and should definitely not be used.

Although you should not try to solder in this way, you should place some solder onto the bit of the iron from time to time, so as to always keep the end nicely covered with some reasonably fresh solder. This is known as 'tinning' the bit incidentally. The point of doing this is that it helps to give a good thermal contact between the bit and the joints, which helps to give good quality results.

It is important to use a suitable soldering iron and type of solder. Any small electric iron should be suitable. By small I mean one that has a rating of somewhere around 15 to 25 watts. There is no need to use something elaborate such as a temperature controlled iron. An ordinary 'no frills' iron will do the job perfectly well. A matching soldering iron stand should be considered an essential extra, not an optional one. For modern electronic work a small bit is required. One of around 2 to 2.5 millimetres in diameter should be suitable. Trying to solder components onto stripboard using an iron fitted with a bit much larger than this can be very difficult indeed. You are likely to find that you have soldered each lead to two or three copper strips!

The solder should be a multi-cored type intended for electrical and electronic work. It is generally available in two gauges. The 22 s.w.g. type is the thinner gauge, and is the most useful for building circuit boards. The much thicker 18 s.w.g. solder is better for larger joints, such as when wiring up the controls and sockets. I would not recommend the 18 s.w.g. type for building circuit boards. If you are only going to buy one gauge of solder, then the 22 s.w.g. type is definitely the one to buy. However, it is useful to buy a small amount of the 18 s.w.g. type as well.

Missing links

If you look at the diagrams which show the component sides of the circuit boards, you will notice that there are numerous pairs of holes joined by lines. These lines represent link wires, which are unavoidable when using stripboard with modern components. These wires can in most cases be made from the pieces of wire trimmed from resistor leadout wires. I do not suggest this method of working as an economy measure – the wire used for most resistor leadout wires is ideal for use as link wires.

In some cases too many link wires might be needed, or some of the wires might be longer than the available leadout trimmings. The link wires must then be made from some 22 or 24 s.w.g. tinned copper wire. 22 s.w.g. is a bit thicker than is ideal, and 24 s.w.g. is slightly thinner than would be ideal. I find that 24 s.w.g. wire is slightly easier to use than 22 s.w.g. wire, but this is just a personal preference.

It is important that the link wires are quite taut, as there is otherwise a risk of short circuits occurring. I find the best way of fitting them is to first solder one end of the link in place. Then thread the other end of the wire through the appropriate hole in the board, pull it tight using some pliers, trim it to length on the underside of the board, and then solder it in place. If you want to make absolutely certain that no short circuits to any link wires occur, simply fit pieces of p.v.c. sleeving over them. However, I have never found it necessary to do this provided the link wires are made reasonably taut.

Wiring-up

In most projects there is a certain amount of wiring from the circuit board to off-board components, such as switches, sockets, potentiometers, etc. There may also be some wiring from one off-board component to another. This is generally known by such names as 'hard wiring', 'spaghetti wiring', and 'point-to-point wiring'. Whatever your preferred term, it is normally completed using thin multi-strand wire. Something like 7/0.2 (i.e. seven cores of 0.2 millimetre diameter wire), p.v.c. insulated wire is suitable. This, or something similar, should be found in any electronics component catalogue. It will probably be described as 'hook-up' wire or 'connecting' wire.

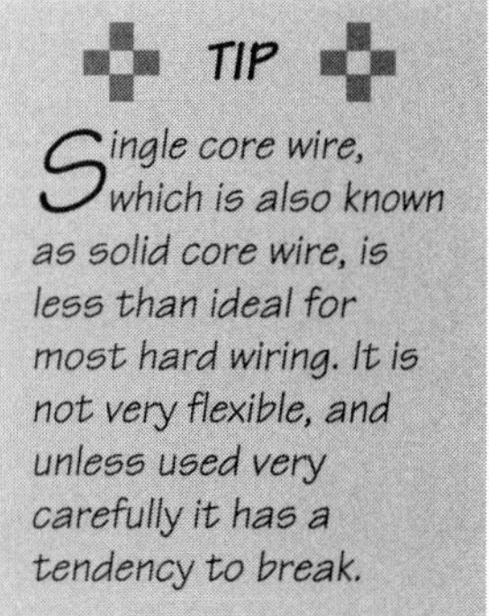

I often use ribbon cable for wiring up projects. Ribbon cable is a multi-way cable that consists of what are effectively several pieces of multi-strand p.v.c. insulated connecting wire laid side-by-side. The wires are fixed together by an overall covering of transparent plastic. It lives up to its name as this type of cable is flat and ribbon-like. It is available as a single coloured (usually grey) cable, or as multi-coloured 'rainbow' cable. The latter is better for wiring-up purposes as having each lead a different colour makes it easy to tell which lead is which. This cable is available from 10 way to about 50 way cable. For our present purposes any ribbon cable will do, but ten way cable is probably the most practical choice.

Screened cable

Audio projects often handle quite low signal levels, which may also be at medium to high impedances. This makes any wiring carry the signals prone to stray pickup of electrical noise such as mains 'hum.' There can also be problems with stray feedback from wiring at the output of the circuit to the wiring at the input. Stray feedback and pickup can be minimised by using screened cables for any wiring that is likely to give problems.

The need for screened wiring can be reduced by housing the projects in metal cases that provide good overall screening of the circuit board and wiring. This does not help reduce stray feedback, but can virtually eliminate unwanted pickup of electrical noise. It is not absolutely essential to use screened leads for any of the wiring in the projects featured in this book, but using screened leads at the input and output of a high gain amplifier or preamplifier greatly reduces the risk of instability due to stray feedback.

A screened lead is basically just an ordinary multi-strand insulated lead, but with a covering of fine wires, and then an overall plastic sheath (Figure 1.6). The fine wires are sometimes woven together (braided), but with most audio screened leads they are simply wrapped around the inner lead. This is known as a 'lapped' cable. Whatever form this wire covering takes, it is called the screen, and is also known as the shield. The easiest way to use most screened audio cables is to use a sharp modelling knife to first remove about 10 to 20 millimetres of the outer sheath. The outer conductor can then be unravelled and twisted to form a thick lead. Some solder can be used to merge the individual fine wires into a single lead that will not fray. A few millimetres of the inner layer of insulation can then be removed using ordinary wire strippers, and the exposed wire is then tinned with solder. There should then be no difficulty in connecting the inner and outer conductors to plugs, sockets, circuit boards, etc.

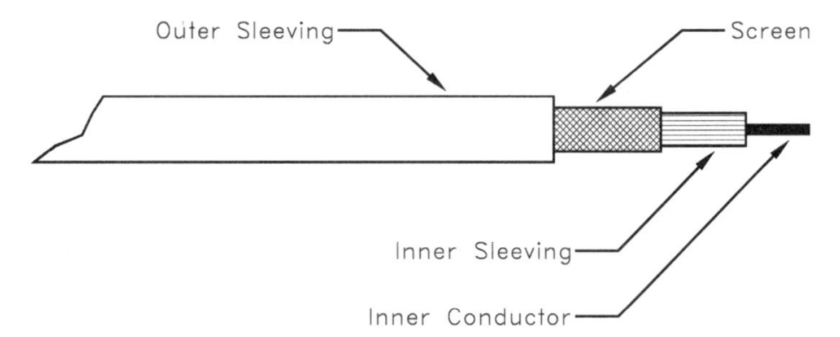

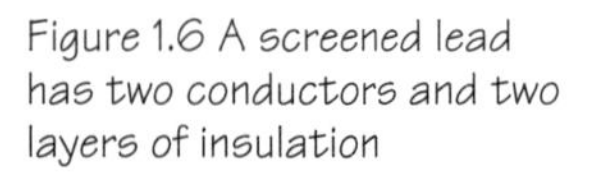
Figure 1.6 A screened lead has two conductors and two layers of insulation

Usually when you are wiring-up a project you will find that there is not just a single wire running from the component panel to an off-board component. In most cases there are two, three or four leads running from the circuit board to each off-board component. There would normally be three leads in the case of a potentiometer for example. In order to help keep things as neat as possible, the circuit boards are designed so that, as far as reasonably possible, all the leads that go to an off-board component emanate from the same area of the board.

When wiring-up a project using ordinary hook-up wire you can make things neater by tying together the wires in each group. Ribbon cable represents an easier way of obtaining the same effect. If you have three leads running from the board to an off-board component, first break away a three way cable from the main piece of ribbon cable, and then cut off a suitable length. This can then be used to provide the three connections to the off-board component. No cable tying is required, as the leads are already held together as a single cable.

A small piece of insulation must be removed from the end of each lead before it can be connected to the tag of an off-board component. It is important to use a pair of wire strippers to do this. You can remove the insulation using scissors, a penknife, etc., but you could easily end up injuring yourself. Also, you are likely to damage the wire, which will then fatigue easily, and is likely to snap before too long. Wire strippers enable the insulation to be removed quickly and easily, and they can be adjusted so that they will cut just the right depth into the insulation, leaving the wires unharmed.

Making connections to off-board components requires a slightly different method of soldering to the one described previously. First, it is important to tin the ends of the leads and the tags of the components with a generous amount of solder. In most cases you will find that the leads and tags take the solder without any difficulty, but some may not. This will be due to dirt or corrosion, which can be scraped away using a miniature file or the blade of a small penknife. Once the tinning has been completed, hook the end of the lead through and around the hole in the appropriate component tag. Then apply the iron and some solder in the normal way, and a good strong joint should be produced. Some component tags are actually pins which lack the hole. With these you simply hook the wire around the pin, and then solder it in place in the normal way.

At the component panel end of the hard wiring you could connect the leads direct to the stripboard. This tends to be a rather awkward way of going about matters, and is also unreliable. The wires tend to break away from the board the pieces of strip to which they are connected.

A better way of tackling things is to use solder pins at the points on the board where connections to off-board components must be made. For 0.1 inch stripboard it is the 1 millimetre diameter pins that are required. There are single and double sided pins, but for the projects featured here you will probably only need the single sided type. These are inserted from the copper side of the board, and pushed home so that very little is left protruding on the underside of the board. A tool for fitting solder pins is available, but they can usually be pushed into place properly with the aid of some pliers. Use a generous amount of solder when connecting the pins to the copper strips, and also tin the tops of the pins with liberal amount of solder. There should then be no difficulty in connecting the leads to them

Mounting tension

There are three basic methods of mounting completed component panels inside cases. The most simple is to mount the board in the guide rails that are moulded into many plastic cases, and a few aluminium types. This is only possible if the board is made a size that will fit into the guide rails with a fair degree of precision. This method is not applicable to most of the projects featured here, although it might be possible in some instances if the board is made larger than is really necessary, so that it will fit the guide rails properly.

My preferred method is to simply bolt the board in place using 6BA nuts and bolts. If your supplier sells only metric sizes, M3 is the nearest metric equivalent to 6BA. Whichever of these you use, 3.3 millimetre diameter mounting holes are suitable. It is important to use spacers about 5 to 10 millimetres long over the mounting bolts, between the case and the component panel. Alternatively, some extra nuts can be used as spacers. If you are using a metal case these will hold the connections on the underside of the board clear of the metal case, so that they do not short circuit through it.

Whatever type of case you use, the spacers will prevent the board from buckling slightly as the mounting nuts are tightened. This buckling occurs because the underside of the board is far from flat once the components have been added, and there are soldered joints protruding on the underside of the board. If you omit the spacers, it is quite possible that the board will buckle so badly that it will crack, or even break into several pieces.

The final method is to use the special plastic stand-offs that are available from most electronic component retailers. These vary somewhat in design, but the most simple type clip into the case and then the board is clipped onto them. The holes in the case and the board must be drilled very accurately if this type of stand-off is to work properly. With the stand-offs of this type that I have used, they never really seemed to hold stripboards in place properly. There is an alternative type which is mounted on the case via self-tapping screws, and these seem to be more reliable. However, with stripboard I much prefer to have the component panels securely bolted to the case.

Layout

There is a vast range of project cases available these days, from inexpensive plastic boxes through to elaborate metal instrument cases which, if used for most projects, would account for more than 50% of the overall cost. There is no one case which is suitable for all the projects in this book. The projects cover various types and complexities, and whereas a cheap plastic box might be appropriate to one project, another might require a larger metal case. Where appropriate, the text gives some guidance as to the best type and size of case for each project. The size of the component panel will give you a good idea of the minimum size of case that is needed.

When designing the overall layout of a project you need to give some thought to how the unit will be used. For instance, do not have sockets too close to, or immediately above control knobs. It is very easy to produce a front panel layout that looks very neat and plausible, but when you plug in the input and output leads the plugs get in the way and make it virtually impossible to adjust some of the controls.

In general it is best to keep inputs well separated from outputs. This is more important with some projects than with others. In the worst cases, having an input socket anywhere near to the output socket is almost certain to render the circuit unstable, possibly preventing it from working at all. Where the layout of a project is critical in this way, or in any other manner, this is pointed out in the text.

The right order

This covers the basics of project construction, and should tell you most of what you need to know. The rest can be learned from looking through one or two large component catalogues, and from experience. A summary of the basic steps in building a project is provided below. This list has the steps in the order that I would suggest you go about things. Not everyone would totally agree with this, and you might prefer to do things your own way in due course, but this gives you a sound initial method of working.

1. Order all the components you need, being careful to get the right ones (including the right types of capacitor etc.).
2. Once you have all the components, identify them all, and check that you have been supplied with the correct parts.
3. Cut the stripboard to size and drill the mounting holes.
4. Fit the components and link wires to the board. Start at one end and work your way methodically to the other side of the board. Fit the solder pins, but do not put the integrated circuits in their holders yet.
5. Work out the case layout, and drill all the holes.
6. Fit the controls and sockets in the case, and then fit the component panel.
7. Wire-up the project using hook-up wire or pieces of ribbon cable.
8. Fit the integrated circuits in their holders.
9. Check the wiring etc., and when all is well, switch on and test the project.

Testing

If a project fails to work, and you do not have the necessary technical skills and equipment to check it properly, all is not lost. The problem could be due to a faulty component, but in all honesty this is highly unlikely provided you buy new components from any respectable supplier. There are many 'bargain packs' of components on offer, and some of these represent outstanding value for money. Others, with the best will in the world, have to be regarded as bags of rubbish. Without technical expertise and some test equipment it is not possible to sort out the good from the bad. Therefore, it is advisable for beginners to only use new components of good quality.

TIP

Try to avoid layouts that result in lots of crossed-over wires when you wire-up the project. This is not just a matter of making the interior of the project look neat. Long leads trailing all over the place can reduce the performance of a circuit or cause instability. Also, if there should be a problem with the project at some time, fault finding will be much easier if the wiring is neat, tidy, and easy to follow.

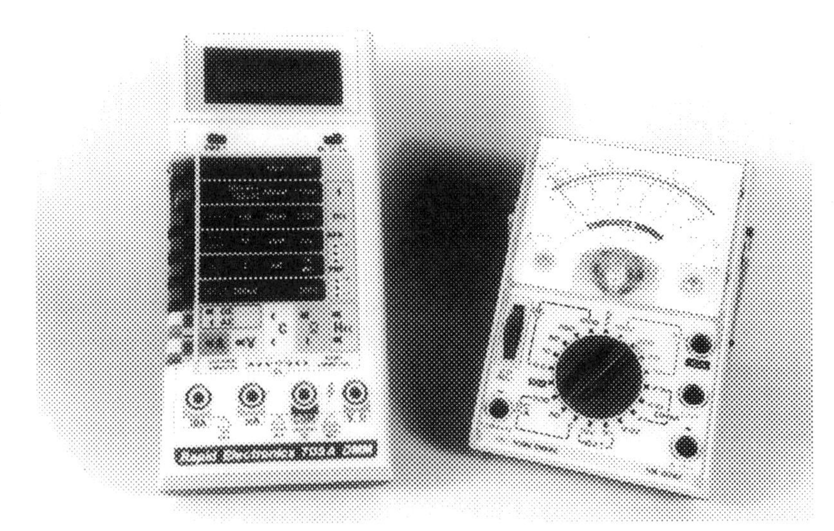

Photo 1.6 A digital multimeter (left) or an analogue type (right) make it much easier to track down simple faults such as bad connections and short-circuits.

Faulty or damaged components

Even if you use top quality components from one of the larger retailers it is obviously still possible that a faulty component could slip through. This is extremely rare though, since components undergo quite stringent testing. Where a faulty component is supplied it is usually due to it having sustained physical damage somewhere along the line. This damage, such as a missing leadout wire, will almost invariably be obvious when you examine the newly obtained components.

At one time there were companies selling semiconductors of dubious quality. Some of these were not quite what they purported to be, while others were so-called 'genuine duds' that failed to work at all. Fortunately, this practice seemed to die out some years ago. Any semiconductors you buy will certainly be the 'real thing' from one of the main semiconductor manufacturers.

Problems are unlikely to be due to faulty components, but it is a good idea to visually inspect the board for signs of damaged components. Are there any discoloured components which overheated when you took too long to solder them into circuit? Are there other signs of damage, such as a polyester capacitor that is parting company with one of its leadout wires? Replace any suspect looking components.

Wrong component?

The wrong component rather than a faulty component is a different matter though. Check the circuit board carefully against the diagrams and the components list, making sure that the right parts are in the right places. With stripboard construction you need to be very careful not to get one or two of the leadout wires connected to the wrong copper strip. Are the electrolytic capacitors and semiconductors all fitted the right way round? You are unlikely to miss out a component, as you would presumably notice that there was one left over. However, it might be worth making a quick check, as small components can sometimes get hidden in the tools and papers on the workbench, and can be overlooked. Link-wires are much more easily missed out. Check that there are the same number of link wires on the component layout diagram and on the circuit board.

Check underside soldering

Probably the most likely area for faults is on the underside of the board. Did you place the solder on the bit of the iron and then transfer it to the joint? This virtually guarantees bad joints, but is the method that many beginners seem to insist on adopting. If there are any suspect joints, remove the solder from them, clean up the copper strip and leadout wire by scraping then with the blade of a penknife, and try again using the correct method of soldering (as described previously in this chapter).

When inspecting the underside of the board look out for joints that are globular in appearance, solder that has a dull rather than a shiny surface, and joints that are covered with burned and black flux. These often mean that the joint concerned is something less than perfect.

Check the breaks

Another point to check is that the breaks in the copper strips are all present and in the right places. If you find that a break has been made in the wrong place, simply solder a piece of wire over the break and make a new one in the right place. Look carefully at each break to determine whether or not it has been fully cut though.

On several occasions I have found that a newly constructed project has failed to work due to a minute trail of copper bridging what should be a break in a copper strip. These are sometimes so thin that they are barely visible with the naked eye. A magnifying glass is more than slightly helpful when inspecting the underside of a stripboard.

Bridged strips?

With stripboard construction the main cause of faults is short circuits between adjacent copper strips due to minute blobs of excess solder. The copper strips are so close together that it is very easy to solder across two strips. This will often be quite obvious, as the strips will be bridged over a length of several millimetres.

Sometimes though, there may be only a very fine trail of solder which is barely visible. Once again, a magnifying glass is very helpful when making a visual inspection of the copper side of the board. Even with an aid to vision, some solder trails can be very difficult to locate. Cleaning the underside of the board with one of the special cleaners that are available can help matters. Look especially carefully at areas of the board where there are a lot of soldered joints.

It helps to bear in mind that if you get all the components in the right place, all the wiring correct, and avoid bad joints and accidental short circuits, the project will work.

TIP

A technique which I have found quite useful is to carefully score between each pair of copper strips using a sharp modelling knife. If there are any solder trails which are so small that they are defeating your eyesight, this treatment should cut through them and remove the problem. Some years ago when I was making large numbers of projects on stripboard I found that this method was successful in curing a surprisingly large percentage of ailing circuit boards.

Down to earth

Many of the projects featured in this book can be built as stand-alone units, or used as part of a larger project. For example, the preamplifiers can be built as external add-ons for existing equipment, or they can be combined with the tone control and power amplifier projects to produce an integrated pre/power amplifier. When adding together several audio projects, especially where there is a large amount of gain through the circuits, remember that there can be problems if the earthing is not spot-on. There can be problems with 'hum' loops, reduced audio quality, or in more severe cases there can even be strong instability. These problems are caused by the resistance in the earth wiring, which will not be very large, but can still result in small voltage drops. Bear in mind that with something like a microphone preamplifier less than one millivolt r.m.s. may be sufficient to fully drive the amplifier. A current flow of one amp through a wire having a resistance of one milliohm is sufficient to produce a voltage drop of one millivolt. It is therefore very easy to produce unwanted feedback from a power amplifier to a preamplifier.

There are two standard approaches to earthing which should avoid any major problems. The most simple of these is the 'spider' earthing system, which simply has each part of the circuit earthed to one earthing point. This method of earthing is shown in Figure 1.7(a). The alternative is the bus-bar method, which is depicted in Figure 1.7(b). Here there is an earth connection that winds its way around the circuit, but not in a random fashion. The earth lead works its way logically through the circuit, starting at the input and working its way through to the output and the power supply. It does not matter which of these systems you use, but you must use one or the other, and not simply arrange the earthing in a higgledy-piggledy fashion.

Figure 1.7 The spider (a) and busbar (b) methods of arranging earth connections

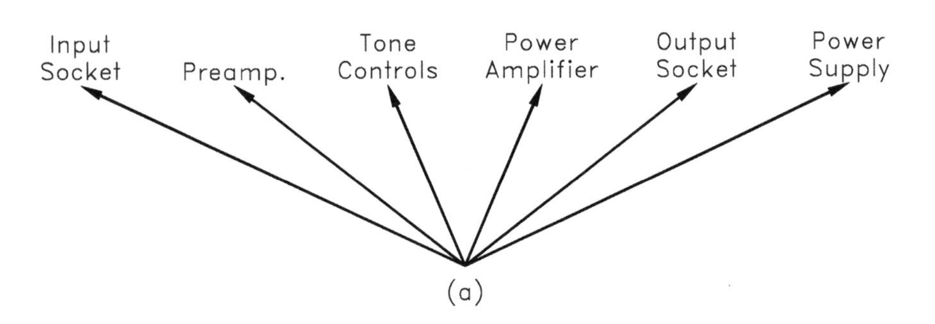

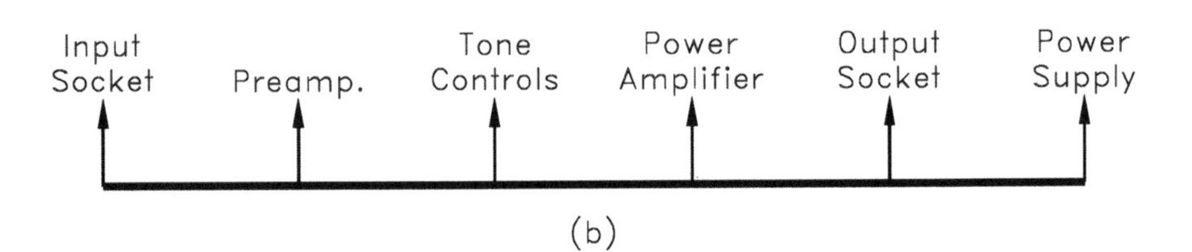

2

Low Z microphone preamplifier

Although some items of audio equipment have a microphone input, it is by no means universal. Simply connecting a microphone to a general purpose audio input such as an auxiliary ('AUX') input is unlikely to give acceptable results. Neither is a 'Tape', 'CD', or 'Tuner' input. These are all high level inputs that are intended for use with a source signal having an amplitude of at least a few hundred millivolts r.m.s. The output level from microphones varies considerably from one type to another, and is also dependent on the exact application and way in which the microphone is used. However, the output level rarely exceeds a few millivolts r.m.s. This is obviously far too low to drive any form of high level input.

Of the popular types of microphone it is the low impedance types that have the lowest output amplitudes. The two common forms of low impedance microphone are dynamic (moving coil) and electret types which do not have a built-in step-up transformer. In both cases the output impedance of the microphone is normally between 100 ohms and one kilohm, although some dynamic microphones have an output impedance of a few kilohms. The output level is generally around 500 microvolts (0.5 millivolts) r.m.s. In order to drive a high level input the signal must therefore be boosted by a factor of about 1000 or so.

The circuit

The circuit diagram for the low impedance microphone preamplifier is shown in Figure 2.1. This is a two stage circuit which uses IC1 in the low noise input stage. Noise performance is an important factor for a preamplifier that will normally have an input level of under one millivolt r.m.s. IC1 is an operational amplifier, but it is a type that is specifically designed for use in low noise, low distortion input stages. The circuit will work using popular devices such as the LF351N and TL071CP, but the noise level will be very much higher. In fact it will probably be more than ten times higher. Over the audio range this circuit achieves an unweighted signal-to-noise ratio of about -76 dB or so (relative to an output level of one volt r.m.s.).

The circuit will work well using other very low noise operational amplifiers, but it is best to use a device which has a bipolar input stage, such as the NE5534AN. This is due to the fact that bipolar input stages give a

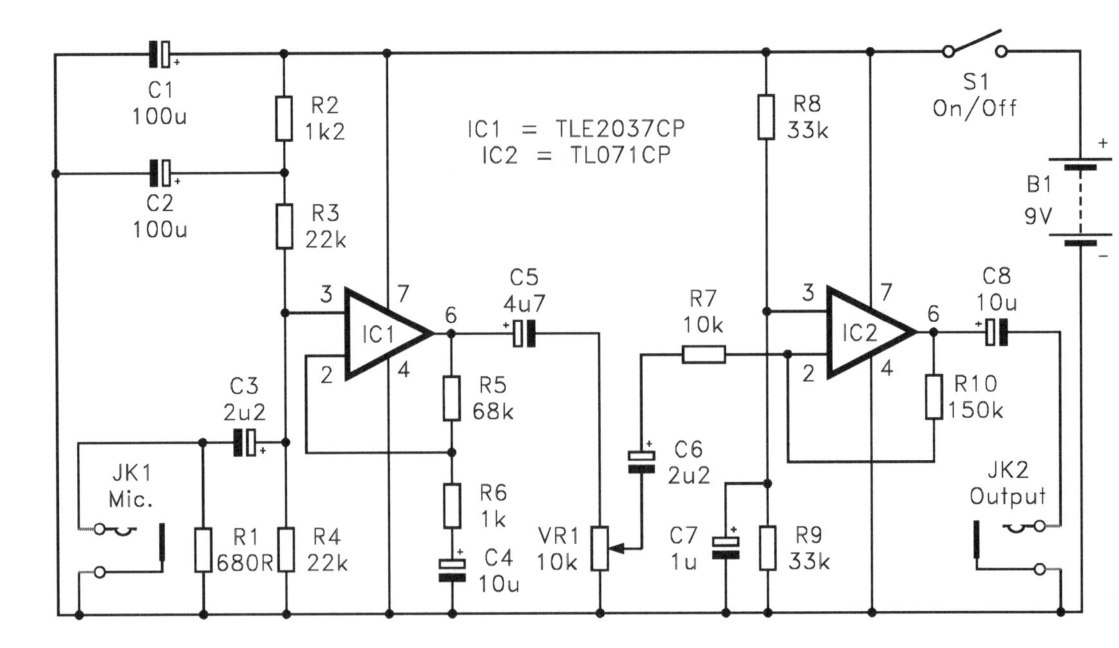

Figure 2.1 Circuit diagram for the low impedance microphone preamplifier

noise level that substantially reduces as the source impedance reduces. Excellent noise performance is therefore obtained in an application such as this where the source impedance is typically just a few hundred ohms. With f.e.t. input devices the noise level is largely independent of the source impedance, which produces an effective reduction in noise performance at low source impedances compared to bipolar devices.

IC1 is used in the non-inverting mode, and its input bias circuit is provided with 'hum' filtering by R2 and C2. R3 and R4 provide the usual bias potential of approximately half the supply voltage, and set the input impedance of the amplifier at 11k. This is somewhat higher than is ideal for this application, and R1 is therefore used to shunt the input impedance to a more suitable figure of about 600 ohms. The values in the bias circuit could be reduced to produce a low input impedance, but this is not a good way of doing things. One problem is simply that a high supply current would be drawn by the bias circuit. Another is that input coupling capacitor C3 would have to be high in value, which would give a large pulse of current through the microphone at switch-on. Over a period of time this could impair the performance of the microphone. This type of input stage is often used to prevent delicate signal sources from being fed with strong pulses of current.

The voltage gain of the input stage is controlled by a negative feedback network comprised of R5 and R6. The voltage gain of the circuit is equal to (R5 + R6) divided by R6, which works out at 69 times in this case. C4 provides D.C. blocking so that the circuit has unity voltage gain at D.C., and the output of IC1 is biased to about half the supply voltage. This enables the highest possible output level to be accommodated before the onset of clipping and serious distortion on the output signal.

C5 couples the output from IC1 to volume control VR1. From here the signal is fed to the second stage of amplification. This utilises an operational amplifier (IC2) in the inverting mode. IC2 is handling a much higher signal level than IC1, making its noise performance less critical. A Bifet operational amplifier such as the TL071CP, TL081CP, and LF351N are therefore adequate for IC2. R8 and R9 bias the non-inverting input of IC2, and C7 provides 'hum' filtering. R7 and R10 are the negative feedback resistors, and the voltage gain of this stage is equal to R10 divided by R7, or 15 times in other words. This gives the circuit an overall voltage gain of up to 1035 times (a little over 60dB). C8 provides D.C. blocking at the output of the amplifier.

The current consumption of the circuit is approximately six milliamps. A small (PP3 size) battery is adequate to supply this, although a higher capacity battery might be more economic if the unit is likely to be used for long periods of time. The circuit can also be powered from a nine or 12 volt battery eliminator, but it is advisable to use a regulated type. An unregulated nine volt battery eliminator might give satisfactory results, but it could well be necessary to increase the values of C2 and C7 in order to obtain a really low 'hum' level on the output signal. There is some advantage in using a higher supply voltage of about 15 to 18 volts. This gives increased dynamic range, and the operational amplifiers provide somewhat better performance with a higher supply voltage. Supply voltages of up to 30 volts are acceptable provided C1 and C2 have a voltage rating that is higher than the selected supply potential.

Construction

The stripboard component layout and hard wiring are shown in Figure 2.2. The underside view of the board appears in Figure 2.3. The board measures 38 holes by 19 copper strips. The principles of stripboard construction are covered in Chapter 1, so refer back to this if you are uncertain about any aspect of constructing the board. Although one or two of the breaks in the copper strips may seem to serve no useful purpose, they are included to reduce stray feedback which could otherwise cause instability. Stray feedback tends to be problematic with stripboard due to the capacitance from one copper strip to the next. Many of the other stripboard designs in this book have breaks in the strips which are included solely to reduce unwanted feedback. In a few cases earthed copper strips are also used to reduce stray feedback. Neither of the integrated circuits are static sensitive, but I would still recommend using holders for them, particularly the TLE2037CP which is not particularly cheap.

The unit will fit into most small to medium size plastic or metal boxes. In audio applications there is some advantage in using a metal case as it helps to shield the circuit from sources of mains 'hum' and other electrical noise. With a highly sensitive circuit such as this the shielding of a metal case becomes virtually essential. When initially testing the uncased prototype board it showed signs of strange low level oscillations on the output signal. It eventually transpired that this was actually stray pick up from a PC situated two or three metres away from the board! Diecast aluminium boxes are probably the best choice for sensitive preamplifiers, but they are relatively expensive. A box of folded aluminium construction makes a good low cost alternative.

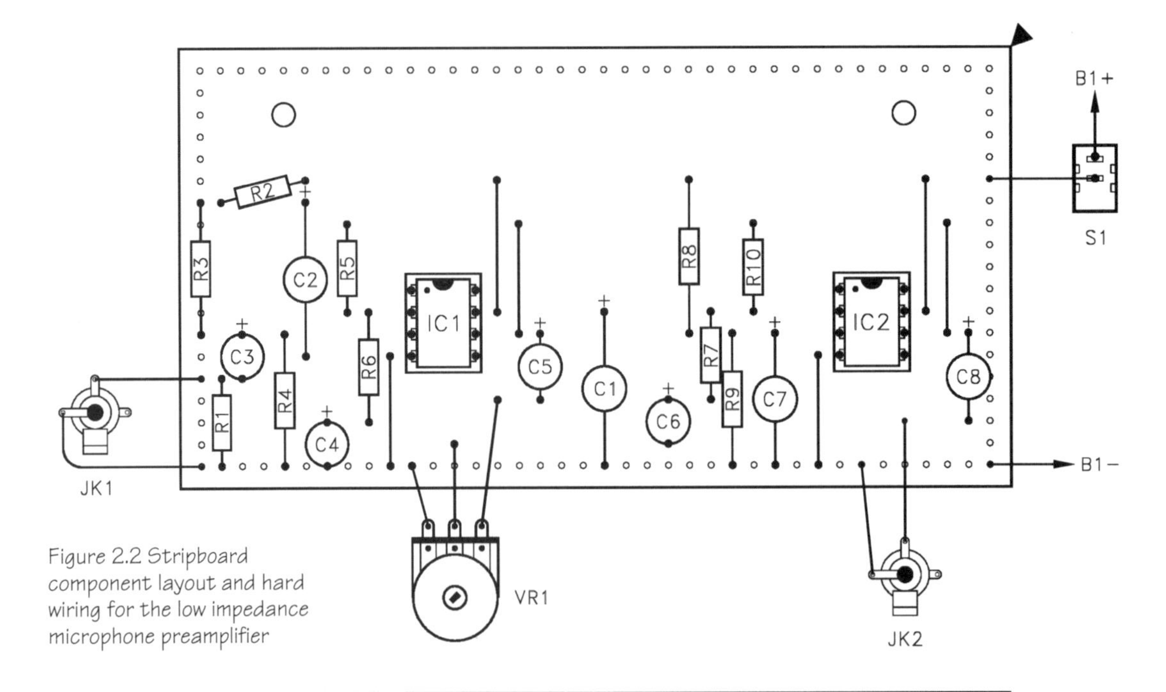

Figure 2.2 Stripboard component layout and hard wiring for the low impedance microphone preamplifier

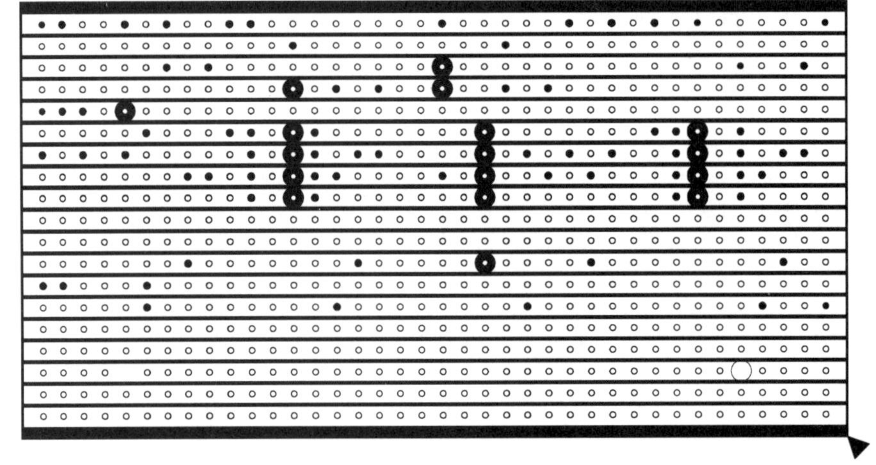

Figure 2.3 Underside view of the board for the low impedance microphone preamplifier

The general layout is not too critical, but with any high gain audio circuit it is prudent to keep the input and output wiring as well separated as possible. It is not essential to use a screened lead to connect JK1 to the circuit board provided this lead is no longer than 75 millimetres. JK2 is a 3.5 millimetre jack socket on the prototype, but a different type of socket (standard jack, phono, etc.) can be used if it will fit in better with your equipment. The normal (open) style 3.5 millimetre jack sockets sometimes cause a certain amount of consternation as they have two tags rather than three. The extra tag connects to a normally closed switch contact. In their original role as earphone sockets, the switch contact was used to switch off the loudspeaker when the earphone was plugged in. In the present application the built-in switch serves no useful purpose, and the extra tag is simply ignored.

In use

Plug the microphone into JK1, and connect JK2 to the amplifier, mixer, or other device via a good quality screened audio lead. Check that VR1 enables the level of the microphone signal to be controlled properly. If it works in reverse (i.e. clockwise rotation reduces the signal level) the connections to the track of VR1 are round the wrong way. In some cases the output level might be too low even with VR1 fully advanced. The gain of the circuit can be boosted somewhat by making R5 higher in value. Try a value of 120k, which will almost double the gain of the circuit.

Components list

Resistors (all 0.25 watt 5% carbon film)

R1	680R
R2	1k2
R3,4	22k (2 off)
R5	68k
R6	1k
R7	10k
R8,9	33k (2 off)
R10	150k

Potentiometer

VR1	110k log carbon

Capacitors

C1,2	100u 16V radial elect (2 off)
C3,6	2u2 50V radial elect (2 off)
C4,8	10u 25V radial elect (2 off)
C5	4u7 50V radial elect
C7	1u 50V radial elect

Semiconductors

IC1	TLE2037CP very low noise op amp
IC2	TL071CP Bifet op amp

Miscellaneous

S1	S.P.S.T. min toggle switch
JK1,2	23.5mm jack socket (2 off)
B1	9 volt (PP3 size)

Metal case, 0.1 inch stripboard measuring 38 by 19 holes, battery connector, 8-pin d.i.l. holder (2 off), control knob, wire, solder, etc.

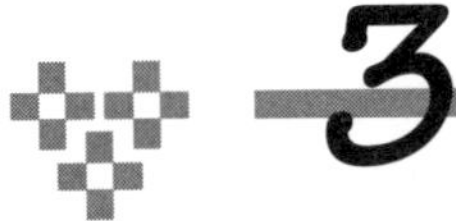

3

High Z microphone preamplifier

High impedance microphones having impedances of around 10k to 50k are mostly of the dynamic and electret varieties. The electret type is basically just an ordinary low impedance electret microphone having a built-in step-up transformer. High impedance dynamic microphones sometimes consist of a low impedance microphone plus step-up transformer, and microphones of this type sometimes offer both high and low impedance outputs. The alternative is to simply have a high impedance coil. With any of these microphones the output characteristic is much the same, with an output impedance of a few tens of kilohms, and an output level of a few millivolts r.m.s. This higher output level and impedance provides what most would regard as a more natural match to the input characteristics of semiconductor amplifiers. However, this does not necessarily translate into a better signal to noise ratio when using a high impedance microphone.

The circuit

The circuit diagram for the high impedance microphone preamplifier appears in Figure 3.1. At first sight this may seem to be identical to the low impedance microphone preamplifier, and it does use exactly the same

Figure 3.1 High impedance microphone preamplifier circuit

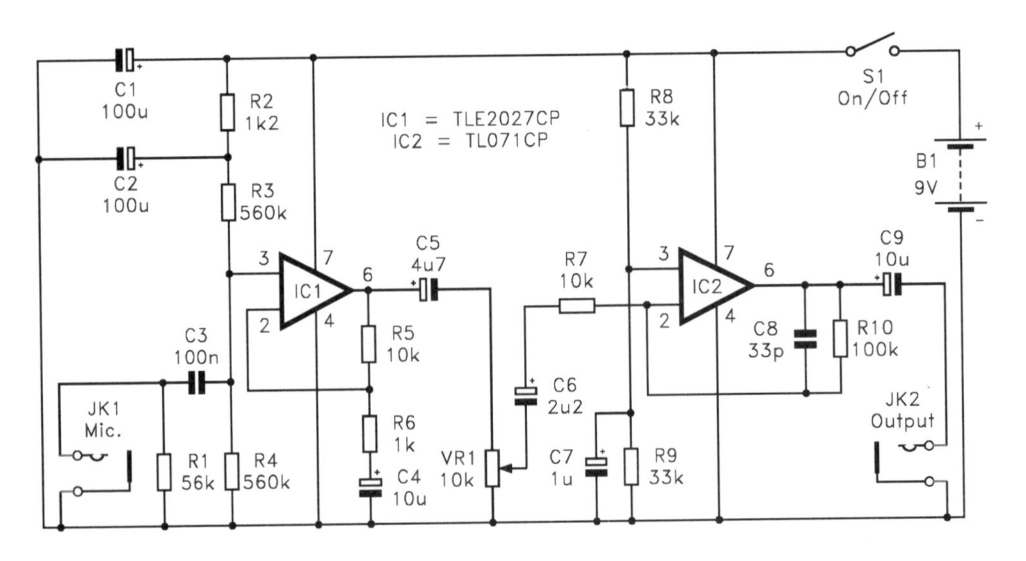

configuration. There are some changes to the circuit values though. R1, R3, and R4 have higher values so that the input impedance is raised to about 47k, which should give good results with any high impedance microphone. The values of R5 and R10 have been reduced in order to reduce the voltage gain of both stages to 10. This gives an overall voltage gain of 100 (40dB), which should be adequate in most cases. However, if necessary the voltage gain of the circuit can be boosted by raising the value of R5 to about 22k.

IC1 is a TLE2037CP in the original circuit, but is a TLE2027CP in this design. These two integrated circuits are virtually identical, and the only difference is that the TLE2027CP has internal compensation for voltage gains of unity or greater, whereas the TLE2037CP is internally compensated for gains of five or more. In theory the TLE2037CP is suitable for use in this circuit, where IC1 is operated at a voltage gain of ten, and it should provide better performance at high audio frequencies. In practice the very wide bandwidth of the circuit makes it difficult to obtain good stability. In order to tame the circuit a TLE2027CP is used for IC1, and the high frequency response of IC2 is rolled off by feedback capacitor C8. If you use an alternative operational amplifier for IC1 it is advisable to select a type which is internally compensated for unity voltage gain, although the circuit seems to work well using an NE5534AN for IC1.

Construction

Figure 3.2 shows the stripboard layout and point-to-point wiring for the high impedance microphone preamplifier. The underside view of the board appears in Figure 3.3. The board has 38 holes by 19 copper strips. The construction notes for the low impedance microphone preamplifier largely apply to this project as well, and will not be duplicated here. Standard jacks rather than the 3.5 millimetre variety have been specified for JK1 and JK2, and this is simply due to the fact that most high impedance microphones seem to be supplied with standard jack plugs.

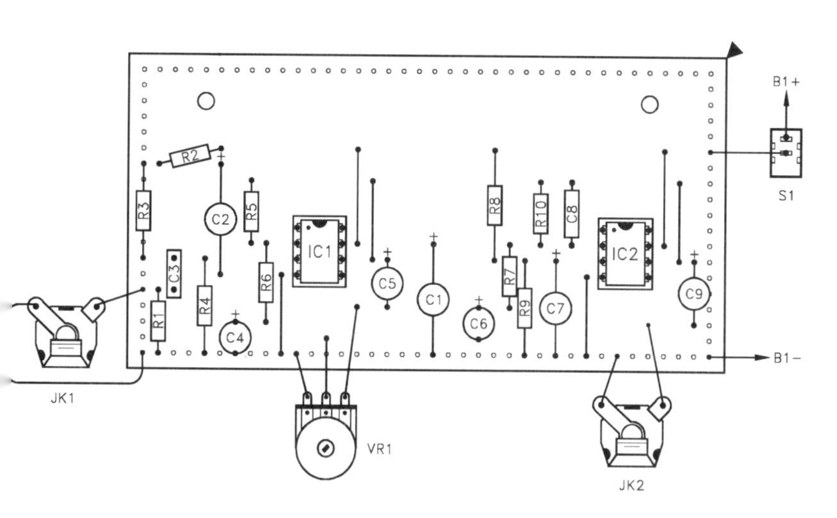

Figure 3.2 Stripboard layout and point-to-point wiring for the high impedance microphone preamplifier circuit

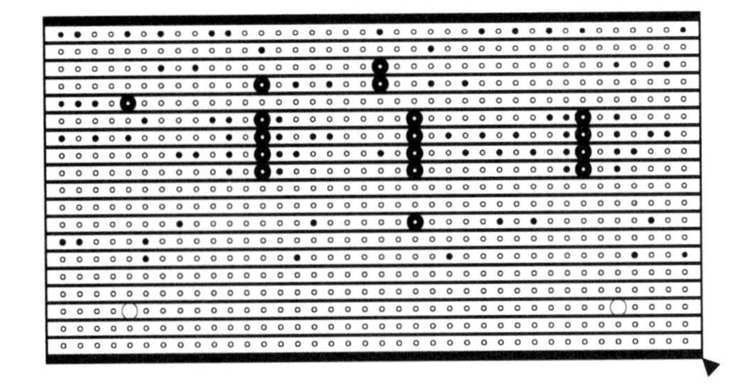

Figure 3.2 Underside view of the high impedance microphone preamplifier circuit board. The board has 38 holes by 19 copper strips

Obviously the input and output sockets can be any type that fits in well with the particular equipment you will be using with the preamplifier.

In use the high impedance microphone preamplifier should provide excellent results. On the face of it the lower voltage gain of this circuit should give it a better signal-to-noise ratio than its low impedance counterpart. To some extent the higher source impedance tends to cancel out the lower gain, giving only a marginal improvement. Over the audio range an unweighted signal-to-noise ratio of around -70dB or so is obtained (relative to an output level of one volt r.m.s.).

Components list

Resistors (all 0.25 watt 5% carbon film)

R1	56k
R2	1k2
R3,4	560k (2 off)
R5	10k
R6	1k
R7	10k
R8,9	33k (2 off)
R10	100k

Potentiometer

VR1	10k log carbon

Capacitors

C1,2	100u 16V radial elect (2 off)
C3	100n polyester (5mm lead spacing)
C4,9	10u 25V radial elect (2 off)
C5	4u7 50V radial elect
C6	2u2 50V radial elect
C7	1u 50V radial elect
C8	33p ceramic plate

Semiconductors

IC1	TLE2027CP very low noise op amp
IC2	TL071CP Bifet op amp

Miscellaneous

S1	S.P.S.T. min toggle switch
JK1,2	Standard 6.35mm jack socket (2 off)
B1	9 volt (PP3 size)

Metal case, 0.1 inch stripboard measuring 38 by 19 holes, battery connector, 8-pin d.i.l. holder (2 off), control knob, wire, solder, etc.

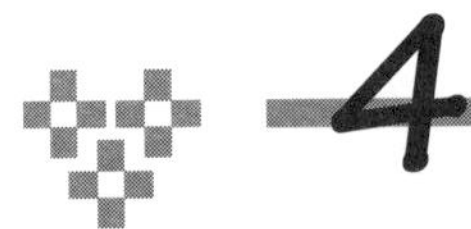

4 Crystal microphone preamplifier

New crystal microphones are something of a rarity these days, although crystal lapel microphones still seem to be manufactured. Crystal microphones were very popular for many years though, and there are still plenty of them in circulation. They are piezo electric devices which operate on the same principle as crystal and ceramic pickups. Also like crystal and ceramic pickups, they must operate into a high input impedance in order to give good results. They provide an output level that is quite high by microphone standards, with signal levels of around 10 millivolts r.m.s. in some cases. An input designed for use with a high impedance dynamic or electret microphone is compatible in this respect. The problem with an input of this type is that its impedance is high in relation to a low impedance microphone input, but is still nothing like high enough to give good results with a crystal microphone. A high impedance microphone input generally has an input impedance of between 10k and 100k, whereas an input impedance of around two megohms is required for a crystal microphone.

Figure 4.1 helps to explain the need for such a high input impedance. If we take the left-hand diagram first, a crystal microphone is effectively a signal source in series with a capacitance that is typically around 10n, but can be anything from about 2.5n to 25n. There is some resistance in series with the signal source as well, since it clearly does not have zero output impedance. However, this resistance is relatively low, and is of little practical significance. The series capacitance is a problem, because it forms a simple highpass filter in conjunction with the input impedance of the amplifier. With an input impedance of around 50k to 100k there is a severe loss of bass response, and with an input impedance of a few kilohms there is a 6dB per octave roll off starting in the middle of the audio range.

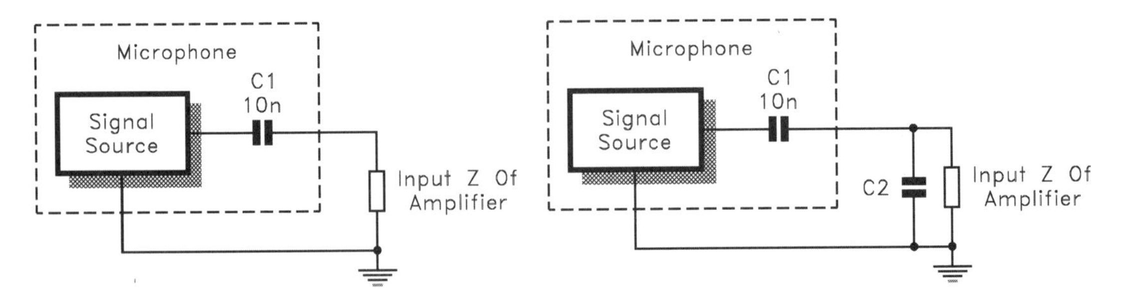

Figure 4.1 Effective circuit of crystal microphone

An input impedance of around two megohms is high enough to ensure that this unwanted highpass filtering only comes into effect at frequencies below the audio range. It is perhaps worth pointing out that long cables do not work well with crystal microphones. The right hand diagram of Figure 4.1 shows the effective circuit when a long cable is used. Here C2 represents the capacitance in the cable, which can be 10n or more for a cable many metres long. The longer the cable the higher the value of C2. C1 and C2 form what is called a capacitive divider. This is much the same as an ordinary potential divider using resistors, but it uses two capacitors, and only operates with A.C. signals. The losses through the divider are governed by the relative values of C1 and C2. The higher the value of C2, the greater the losses. Although one might expect the losses to increase at high frequencies, there is no high frequency loss through the divider itself. However, the impedance of a capacitive divider drops at high frequencies, causing increased loading on the signal source. This can result is some loss of high frequency response.

The practical consequence of this is that the output level from a crystal microphone effectively decreases if a long cable is used, and there might also be some loss of treble response. The longer the cable, the lower the output signal. Most modern screened audio cables have reasonably low capacitance figures (typically about 350p per metre), but this still precludes the use of long leads with crystal microphones unless a preamplifier is used. Connect the microphone to the preamplifier by a lead of no more than a few metres in length, and then connect the output of the preamplifier to the amplifier, tape recorder, etc. via the long cable.

The circuit

Figure 4.2 shows the circuit diagram for the crystal microphone preamplifier. This is just a slight variation on the high impedance microphone preamplifier described previously. R4 has been reduced in value so that the

Figure 4.2 Circuit diagram for the crystal microphone preamplifier

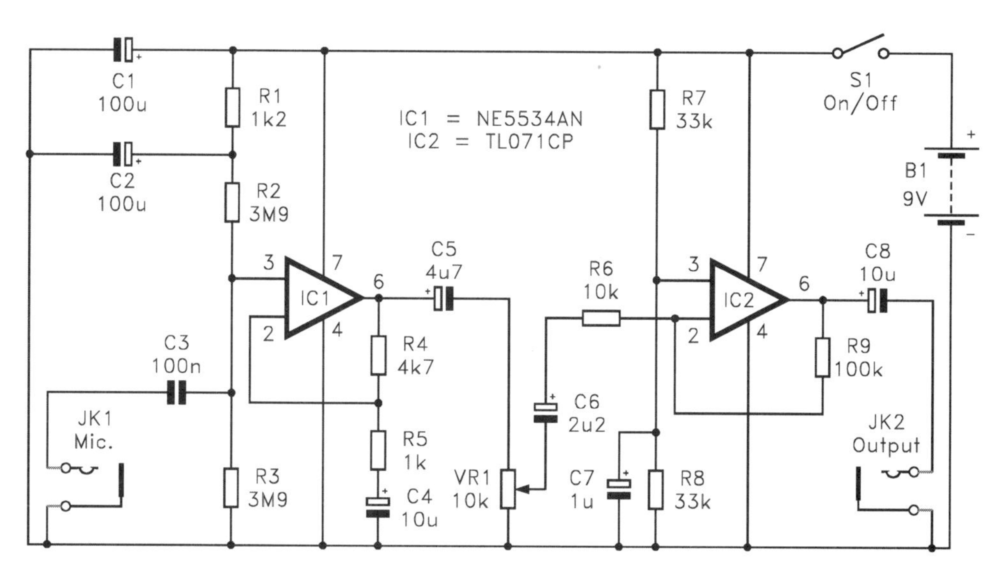

maximum voltage gain of the circuit is reduced slightly. The closed loop voltage gain of IC1 is 5.7, and that of IC2 remains at 10, giving an overall voltage gain of 57. This is adequate for most crystal microphones, but some 'stick' types have relatively low output levels, and might require more gain. This can be achieved by increasing the value of R4 to about 12k.

The input impedance of the amplifier is set at 1.95M by input bias resistors R2 and R3. This should be high enough to give a good bass response from any crystal microphone. In the prototype I used an NE5534AN for IC1 rather than a TLE2027CP, simply because I happened to have an NE5534A to hand at the time. The circuit will work well with either type. The circuit was found to be stable without introducing any high frequency roll-off to IC2.

Construction

Details of the stripboard layout, hard wiring, and cuts in the copper strips are provided in Figures 4.3 and 4.4. The stripboard measures 38 holes by 19 copper strips. Construction of this project follows along exactly the

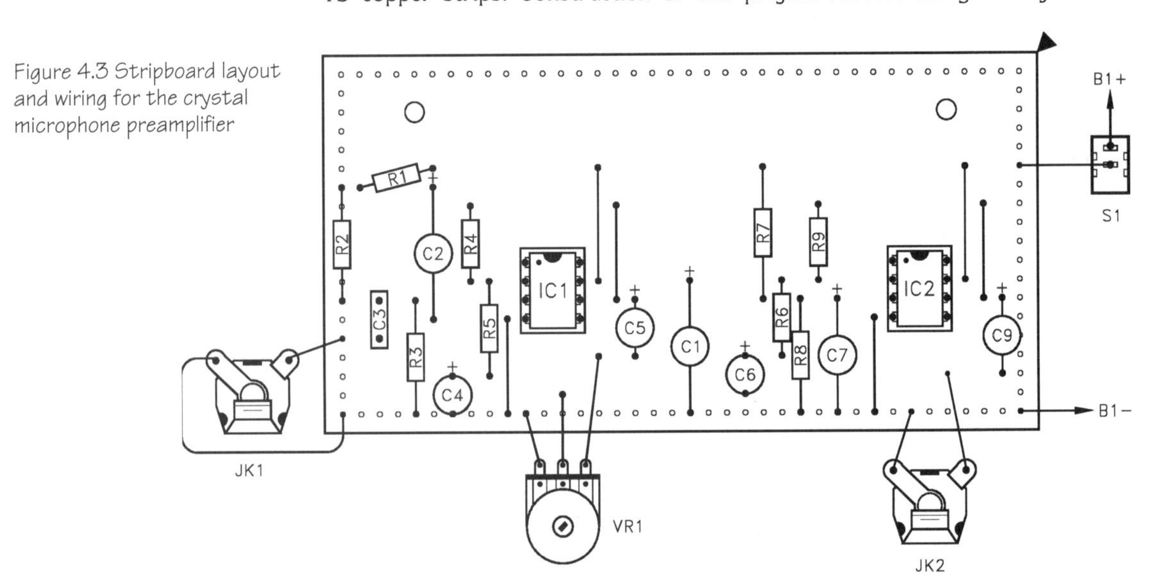

Figure 4.3 Stripboard layout and wiring for the crystal microphone preamplifier

Figure 4.4 Underside view of the crystal microphone preamplifier

same lines as construction of the high impedance microphone preamplifier. However, bear in mind that the very high input impedance of this circuit makes it extremely vulnerable to stray pickup of electrical noise. It is essential to use a metal case that has good screening properties, and the microphone cable must be a high quality type. Otherwise the slight background 'hiss' of the preamplifier will be drowned in mains 'hum', radio frequency breakthrough, and other electrical noise!

Note that there is a version of the NE5534 which lacks the 'A' suffix. The NE5534A has a superior specification in certain respects, including a lower audio noise level. An NE5534N or NE5534P will work quite well in this circuit though.

Components list

Resistors (all 0.25 watt 5% carbon film)

R1	1k2
R2,3	3M9 (2 off)
R4	4k7
R5	1k
R6	10k
R7,8	33k (2 off)
R9	100k

Potentiometer

VR1	10k log carbon

Capacitors

C1,2	100u 16V radial elect (2 off)
C3	100n polyester (5mm lead spacing)
C4,8	10u 25V radial elect (2 off)
C5	4u7 50V radial elect
C6	2u2 50V radial elect
C7	1u 50V radial elect

Semiconductors

IC1	NE5534AN very low noise op amp
IC2	TL071CP Bifet op amp

Miscellaneous

S1	S.P.S.T. min toggle switch
JK1,2	Standard 6.35mm jack socket (2 off)
B1	9 volt (PP3 size)

Metal case, 0.1 inch stripboard measuring 38 by 19 holes, battery connector, 8-pin d.i.l. holder (2 off), control knob, wire, solder, etc.

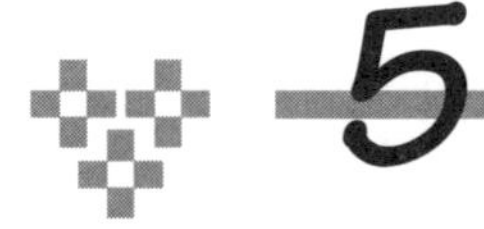

5

Guitar and general purpose preamplifier

The output levels provided by guitar pickups varies considerably from one to another. In general, the more recent and expensive types have quite high output levels, while the older and cheaper types often provide relatively modest signal levels. High output guitars can be used with most power amplifiers without any problems, since they do not require an input specifically for guitars. A high output pickup provides a strong enough signal to drive any high level input, such as an 'Aux.', 'Tuner', or 'Tape' input.

The same is not true of low and medium output types, which usually provide only quite low volume levels when used with high level inputs. On the face of it an RIAA cartridge input should be suitable, with an input impedance of about 50k and only about 5 millivolts r.m.s. needed to fully drive the amplifier. Unfortunately, an RIAA input has equalisation that provides substantial bass boost and treble cut. You may well get adequate volume, but the guitar will sound pretty strange.

This preamplifier is primarily intended for use when using a low output guitar with an insensitive amplifier, but it can also be used as a general purpose preamplifier for use whenever a signal boost is required. For example, a fair percentage of older radio tuners, cassette decks, etc. do not fully drive some modern amplifiers. The circuit provides a voltage gain of 13, but this is easily altered to suit individual requirements, as is its input impedance of about 47k.

The circuit

The circuit diagram for the guitar preamplifier appears in Figure 5.1. A simple single stage circuit is all that it needed in an application like this, where only a low voltage gain is required. The circuit is basically just an operational amplifier used in the standard non-inverting mode. R2 and R3 bias the non-inverting input of IC1, and set the input impedance at 280k. This is shunted by R1, giving an overall input impedance of about 47k, which should give good results with any normal type of guitar pickup. The closed loop voltage gain is controlled by R4 and R5, and is equal to (R4 + R5)/R5, or just over 13 with the specified values. If some extra voltage gain is needed, the value of R4 can be increased to about 100k, which boosts the voltage gain to around 27. Within reason, the input impedance

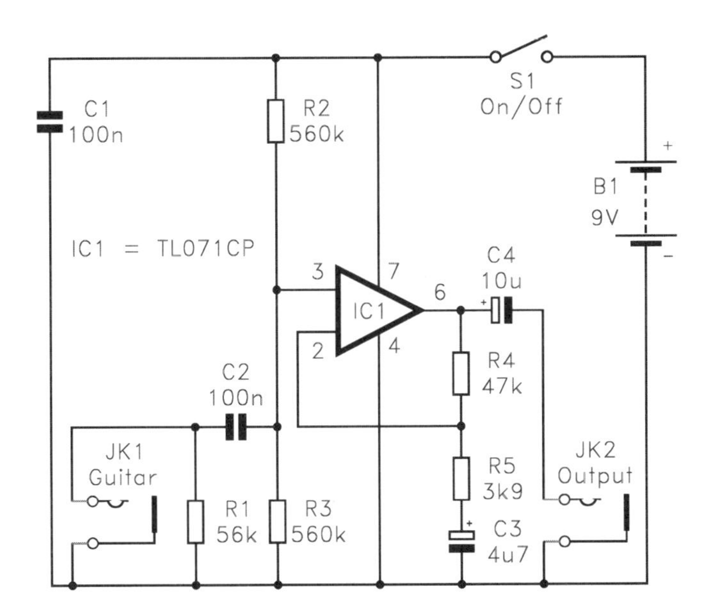

Figure 5.1 Circuit diagram for the guitar preamplifier

of the circuit can be changed by altering the value of R1. The input impedance of the circuit is equal to the value of R1 in parallel with 280k.

Although the input signal will not be at a particularly high level, the use of a very low noise operational amplifier is not really justified. A Bifet device such as the TL071CP or LF351N gives adequate noise performance for most purposes. A device such as the TLE2027CP or NE5534AN can be used if a very low noise level is deemed important, but there could be problems with instability using 'superior' operational amplifiers. The current consumption of the circuit is about two milliamps.

Construction

Refer to Figure 5.2 for the stripboard layout and wiring for this project. The underside view of the circuit board appears in Figure 5.3. The board measures 20 holes by 18 copper strips. Construction of this project is

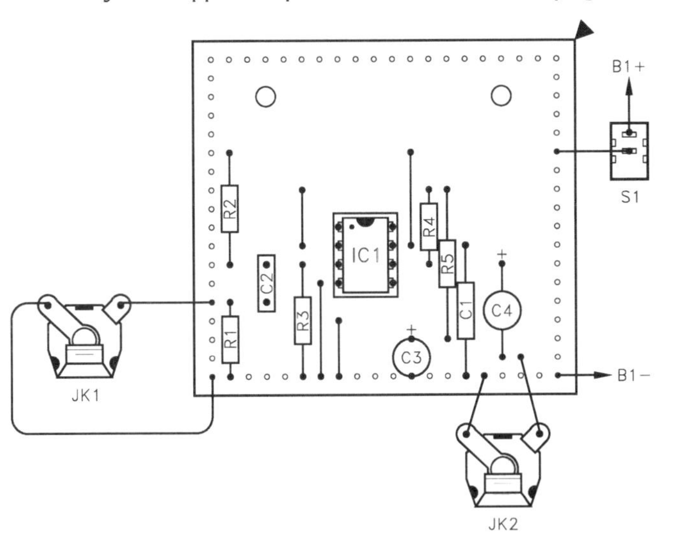

Figure 5.2 Stripboard layout and wiring for the guitar preamplifier

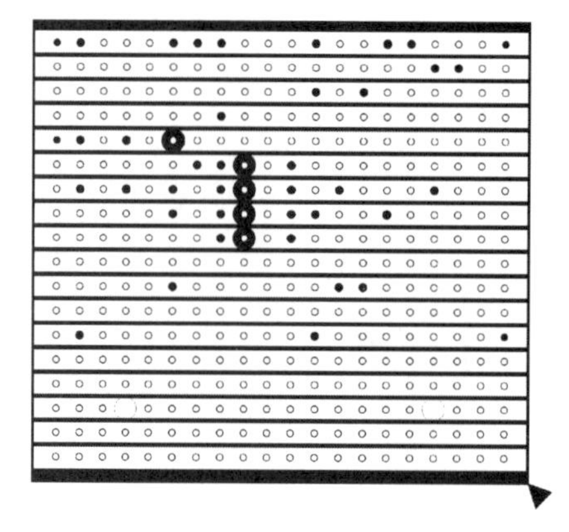

Figure 5.3 Underside view of the circuit board

very straightforward, and offers nothing out of the ordinary, but bear in mind the points about screening sensitive audio projects that were mentioned in the chapter dealing with the low impedance microphone preamplifier.

Components list

Standard jack sockets are the most appropriate type to use if the unit is used as a guitar preamplifier, but it might be better to use an alternative type if the preamplifier is used in a different role.

Resistors (all 0.25 watt 5% carbon film)

R1	56k
R2,3	560k (2 off)
R4	47k
R5	3k9

Capacitors

C1	100n ceramic
C2	100n polyester (5mm lead spacing)
C3	4u7 50V radial elect
C4	10u 25V radial elect

Semiconductor

IC1	TL071CP Bifet op amp (or similar)

Miscellaneous

JK1,2	Standard jack socket (2 off)
B1	9 volt (PP3 size)
S1	S.P.S.T. min toggle

Metal case, 0.1 inch pitch stripboard having 20 holes by 18 copper strips, battery connector, 8-pin d.i.l. holder, wire, solder, etc.

6 Scratch filter

In these days of digital audio a scratch filter is of rather less use than it once was. However, those who have old and cherished recordings on L.P.s and 78s can still put this type of audio processor to good use. Also, 'historic' recordings on compact disc sometimes have little or no audio processing applied during the recording process, and can benefit from some filtering.

It is important to realise that a scratch filter is good at combatting small scratches that only contain high frequencies (usually termed 'surface noise'), but is relatively ineffective at counteracting larger scratches. In fact simple filtering is unlikely to give a worthwhile improvement with large scratches. Large blemishes in an audio signal can only be effectively counteracted using a sophisticated noise blanker or complex digital processing. It is also important to realise that a scratch filter will attenuate all high frequency components in the signal. Some of the wanted signal will therefore be lost along with the surface noise. The degree to which this affects the audio quality depends on the high frequency content in the recording. With a comparatively modern record there will generally be a noticeable loss of 'brightness', but with older recordings there may be little high frequency content to lose.

This type of filtering can also be used to combat 'hiss' type noise. Modern electronics has greatly reduced problems with noise of this type, but it can still trouble stereo f.m. reception in poor signal areas, and is sometimes quite noticeable on older cassette recordings. Bear in mind though, that a dynamic noise limiter gives superior results when 'hiss' type noise is a problem, although a unit of this type is inevitably more complex and expensive than a simple filter.

All a scratch filter does is to apply some lowpass filtering which attenuates the higher audio frequencies. The optimum cut-off frequency is to some extent a matter of opinion, and also depends on the size of the scratches on the record. With minute scratches the frequencies generated are at very high audio frequencies and beyond. Large scratches are likely to produce strong signals at much lower frequencies as well, possibly down into the middle audio range.

The circuit

The circuit diagram for the scratch filter appears in Figure 6.1. IC1 merely acts as an input buffer stage which ensures that the filter circuit is fed from a suitably low source impedance. It provides the circuit with an input impedance of 50k. There is no generally agreed standard for the best cut-off frequency and attenuation rate for a scratch filter. However, the – 6dB point is generally around 5kHz to 7kHz. A high attenuation rate has the advantage of a large amount of noise reduction without having to

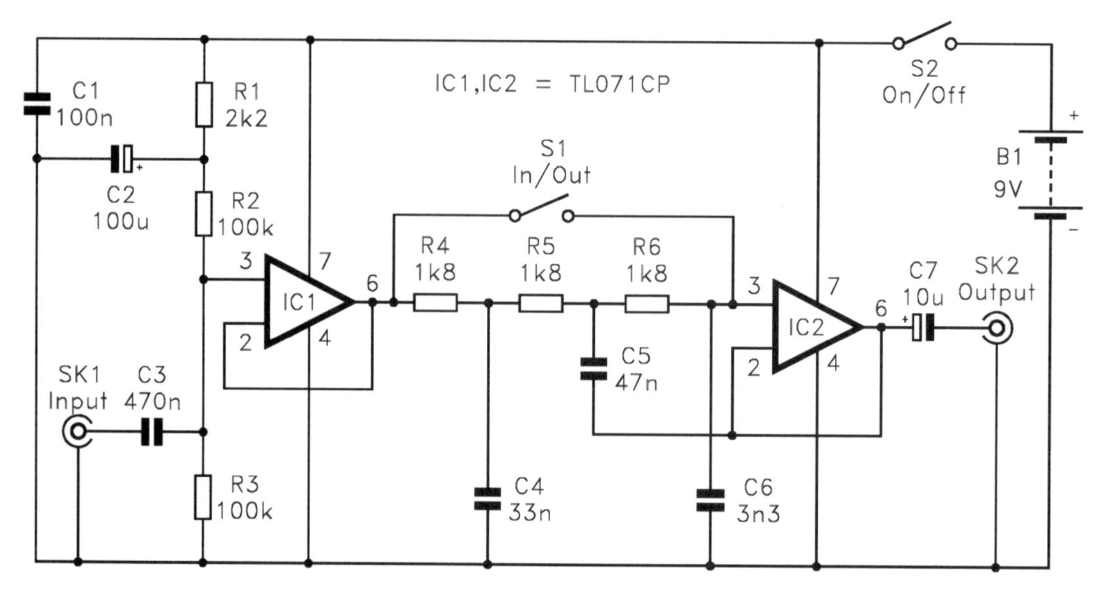

Figure 6.1 The scratch filter circuit diagram

set a low cut-off frequency. This helps to minimise the loss of audio quality. On the other hand, many people feel that this 'brickwall' filtering is rather too obvious and intrusive. A compromise approach has been used with this filter, which has a -6dB point at 6kHz, and an ultimate attenuation rate of 18dB per octave. The frequency response of the circuit is shown in Figure 6.2. It is possible to obtain a slightly flatter response below the cut-off frequency, but component tolerances have to be tighter in order to avoid a resonant peak immediately below the cut-off frequency. There is also a slight risk of the circuit breaking into oscillation. This response is conservative but safe, and gives good results in practice.

Returning to the filter circuit, this is a conventional three stage active circuit having IC2 as the buffer amplifier. A multi-stage active filter offers improved performance over an equivalent passive type due to the use of one or more 'bootstrapping' components. In this case C5 is the 'bootstrapped' component. Explanations of active filters often seem to involve pages of mathematics, but here we will consider things in simple terms. Within the passband of the circuit C5 has no effect, since any change in

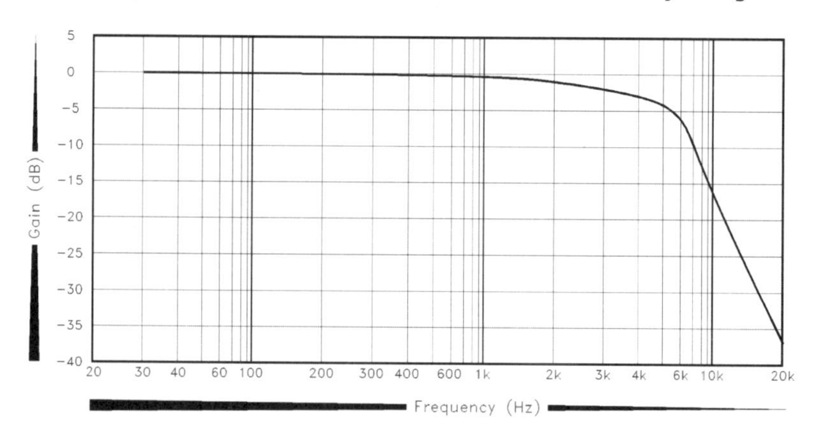

Figure 6.2 The frequency response for the scratch filter

the voltage at the input end of this component is precisely matched by a change at the output end. Outside the passband the signal level at the input end of C5 will not be matched by the signal at the output of IC2, and C5 then becomes part of the filter network. This adding in of C5 as the attenuation rises provides a much more rapid introduction of the filtering, and avoids any major loss of signal below the cut-off frequency.

S1 provides a simple but effective means of switching out the filtering when it is not required. It simply bypasses the filter resistors so that the output of IC1 directly drives the non-inverting input of IC2. The output impedance of IC1 forms a lowpass filter in conjunction with C6, but the cut-off frequency is too high to significantly affect the audio frequency performance of the circuit. As there is no significant voltage drop through R4 to R6, this method of in/out switching is almost totally free from switching 'clicks.' If preferred, the method of in/out switching shown in Figure 6.3 can be used. This simply bypasses the filter circuit altogether when set to the 'out' position.

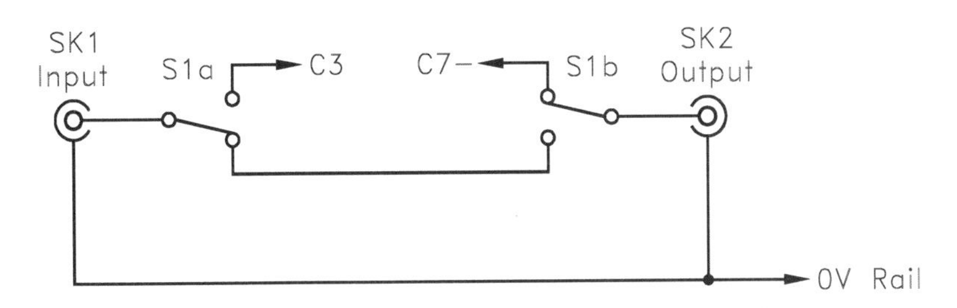

Figure 6.3 An alternative method of in/out switching

If desired, the cut-off frequency of the filter is easily altered. The cut-off frequency is inversely proportional to the value of R4, R5, and R6. For example, increasing the value of these resistors from 1k8 to 2k0 is an increase of just over 10%, and would therefore reduce the cut-off frequency by a little over 10%. This would bring the -6dB point from 6kHz down to around 5.3kHz. All three resistors should have the same value.

The current consumption of the circuit is about four milliamps. R1 and C2 provide 'hum' filtering in the bias circuit, and the unit should therefore work well from a nine volt battery eliminator, even if it is not a type which has a well smoothed and regulated output.

Construction

The stripboard layout and wiring for the scratch filter appear in Figure 6.4. The underside view of the board is shown in Figure 6.5. The board measures 42 holes by 14 copper strips. This is another simple project which should be easy to construct. SK1 and SK2 are phono sockets on the prototype, and this is now the type that is most common on hi-fi equipment. They can obviously be changed to another type of audio connector if phono types are inappropriate for your particular set-up. As the unit is designed to work with a high level signal, and it has only unity voltage gain, it is not necessary to use screened leads to connect the circuit board

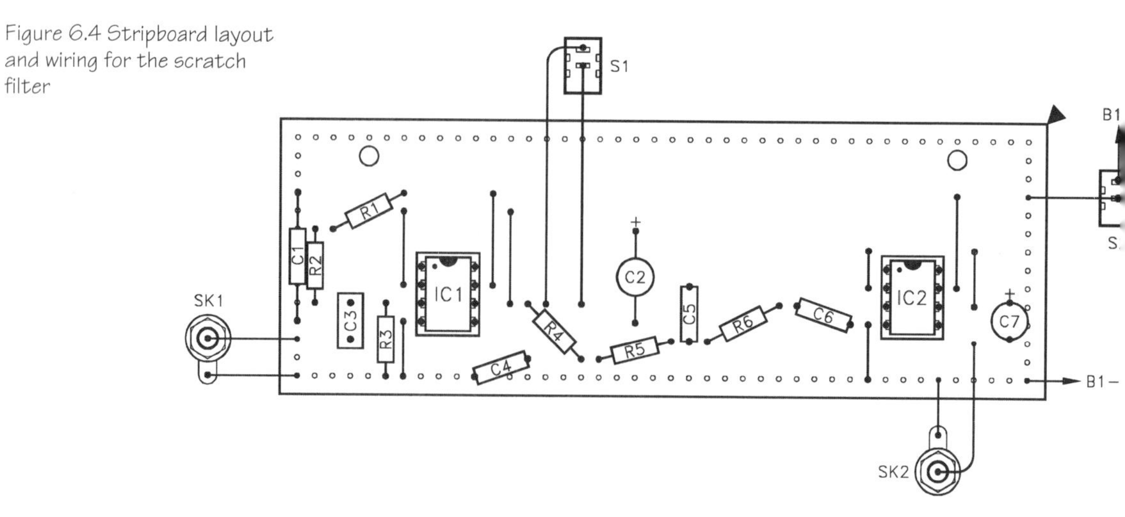

Figure 6.4 Stripboard layout and wiring for the scratch filter

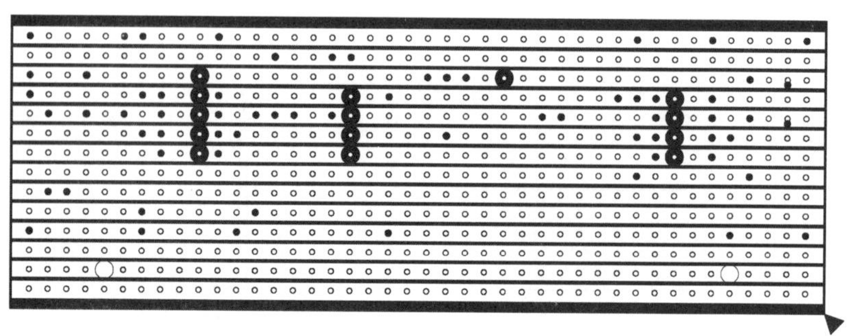

Figure 6.5 Underside of the scratch filter board

to SK1 and SK2. I would still recommend the use of a metal case to screen the circuit, and the external input and output leads must be screened types.

If a stereo version of the unit is required, two filters will be needed, one for each stereo channel. Either build two circuit boards, or use a double size piece of board (i.e. 42 holes by 28 strips) so that two circuits can be built onto the one board. S1 should be a double pole switch, with one pole used in each channel. S2 remains a single pole switch, with its lower tag connecting to the apposite solder-pin of both filters.

In use

The filter could be connected between a magnetic cartridge and the input of the RIAA preamplifier, but this will not give the best signal-to-noise ratio. It would also render the unit very sensitive to pickup of mains 'hum.' It is better if it can be connected between the preamplifier and the power amplifier. With integrated preamplifiers and power amplifiers this can be difficult, but preamplifier outputs and power amplifier inputs are sometimes provided. Alternatively, there may be a tape monitor facility or something of this nature that can be utilised. A facility of this type will almost certainly provide a means of switching in the filter or bypassing it, making S1 in the filter unnecessary. Another option is to build an RIAA preamplifier into the scratch filter. The output of the preamplifier is then used to drive the input of the scratch filter. The output of the filter can then be connected to any high level input of the amplifier ('Tape', 'CD', 'Aux.', etc.). An RIAA preamplifier design is featured elsewhere in this book. The effect of the filter should be very obvious when the unit is used with any recording that suffers from even a modest amount of surface noise.

Components list

Resistors (all 5% 0.25 watt carbon film)

R1	2k2
R2,3	100k (2 off)
R4,5,6	1k8 (3 off)

Capacitors

C1	100n ceramic
C2	100u 16V radial elect
C3	470n polyester (5mm lead spacing)
C4	33n polyester (7.5mm lead spacing)
C5	47n polyester (7.5mm lead spacing)
C6	3n3 polyester (7.5mm lead spacing)
C7	10u 25V radial elect

Semiconductors

IC1,2	TL071CP Bifet op amp, or similar (2 off)

Miscellaneous

SK1,2	Phono sockets (2 off)
S1,S2	S.P.S.T. min toggle (2 off)
B1	9 volt (PP3 size)

Metal case, 0.1 inch pitch stripboard measuring 42 holes by 14 copper strips, battery connector, 8-pin d.i.l. holder (2 off), wire, solder, etc. Two sets of components are required for a stereo scratch filter, apart from the case, battery, S2, and S1. The latter should be a D.P.S.T. min toggle switch.

7 Rumble filter

Very old recordings often seem to be contaminated with rumblings at relatively high frequencies, and need a higher cut-off frequency to achieve good noise reduction. A higher cut-off frequency does not necessarily have a disastrous outcome for the audio quality, as 'ancient' recordings are often lacking in bass content anyway.

A rumble filter is another device that is designed to combat noise on old disc recordings. It operates at the other end of the spectrum to a scratch filter, and provides highpass filtering that attenuates low frequency signals. This noise can be caused by low frequency mechanical sounds that originate in the record deck itself, and find their way to the stylus via the turntable and the record. With a good quality deck any noise of this type should be at a very low level though. These days the main cause of low frequency noise tends to be imperfections in the records themselves. These can be due to moulding faults, or the record becoming warped. In the case of 'historic' recordings there are usually a good selection of low frequency sounds, which seem to be a mixture of noise caused by imperfections in the original disc itself, and mechanical noise originating from the original recording apparatus.

Simple filtering can be very effective at reducing 'rumble', regardless of its cause. Like scratch filtering, it will also tend to remove some of the wanted signal. The cut-off frequency should be the lowest one that will give the required degree of noise reduction, so that the affect on the main signal is minimised. In practice the ideal cut-off frequency varies from one recording to another. With recent records the low frequency noise is generally at very low frequencies, and may be largely in the infra-audio range (i.e. below the lowest audio frequency). A very low cut-off frequency will then deal with the noise, while having a minimal affect on the main signal.

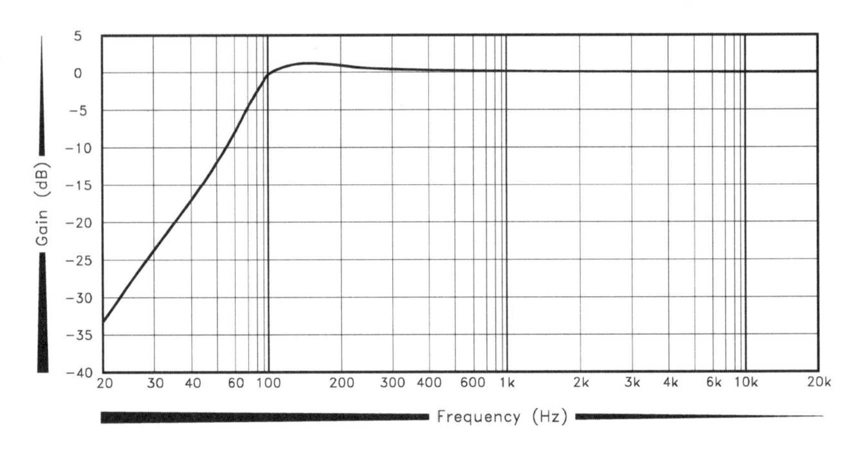

Figure 7.1 Frequency response of the rumble filter

Figure 7.1 shows the frequency response for this rumble filter, which has an ultimate attenuation rate of 24dB per octave. The roll-off commences at a frequency of 100Hz, with a -6dB point at about 65Hz, and well over 30dB of attenuation at the lowest audio frequency of 20Hz. With some programme sources a somewhat lower cut-off frequency would be desirable, and the circuit is easily modified to provide alternative cut-off frequencies.

The circuit

The circuit diagram for the rumble filter appears in Figure 7.2. IC1 acts as a buffer stage at the input, and this stage is identical to the input stage of the scratch filter. The filter is an active fourth order type, and this has similarities to the scratch filter circuit. The resistors and capacitors have been swapped over so that a highpass action rather than a lowpass type is obtained, and there is an extra filter stage including a second 'boot-strapped' component. This gives an ultimate attenuation rate of 24dB per octave, as opposed to the 18dB per octave roll-off of the scratch filter. There is no D.C. path from the output of IC1 to the non-inverting input of IC2. Accordingly, R7 and R8 provide biasing for IC2, and their parallel resistance forms part of the filter network.

S1 enables the filtering to be switched out, and it simply bypasses the filter components when closed. Due to the component tolerances it is possible for a small D.C. offset to be produced across S1. This could produce 'clicks' when S1 is opened and closed. To minimise this problem R2, R3, R7, and R8 should have a tolerance of 1% or better. If preferred, the filtering can be switched out using a d.p.d.t. bypass switch (see Figure 6.3). The current consumption of the circuit is about four milliamps.

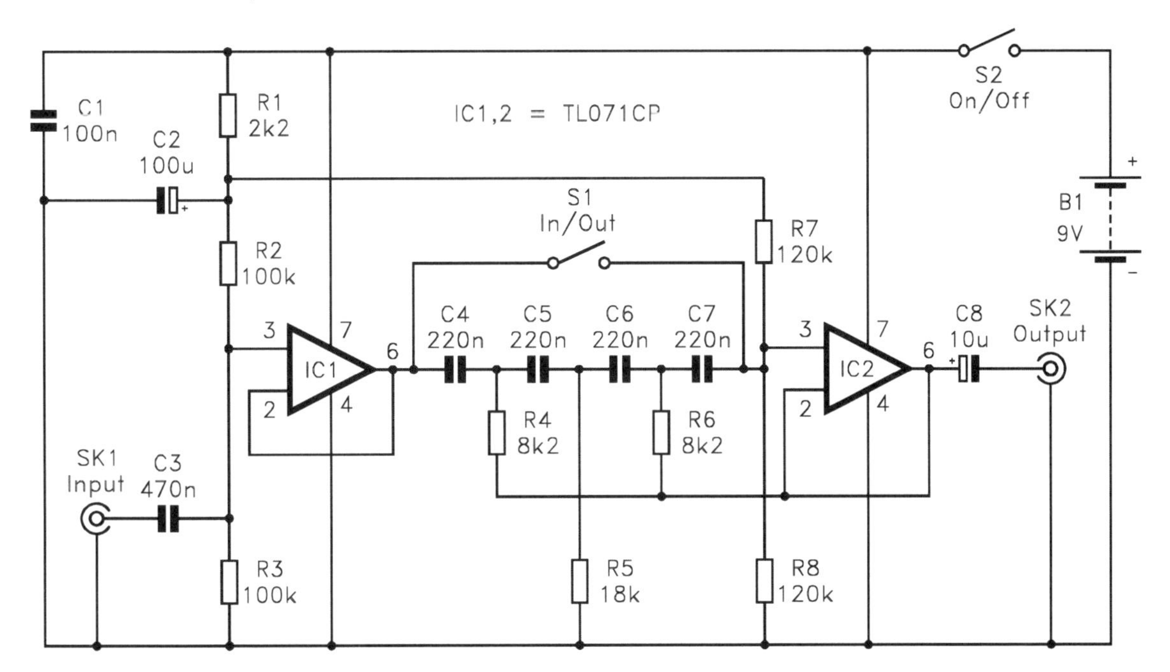

Figure 7.2 Rumble filter circuit

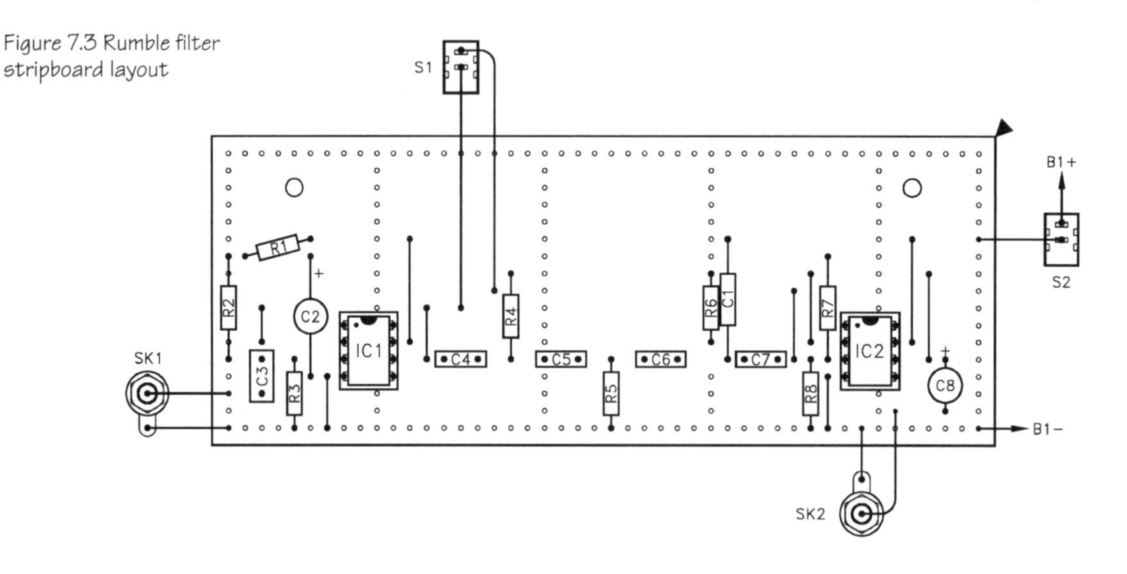

Figure 7.3 Rumble filter stripboard layout

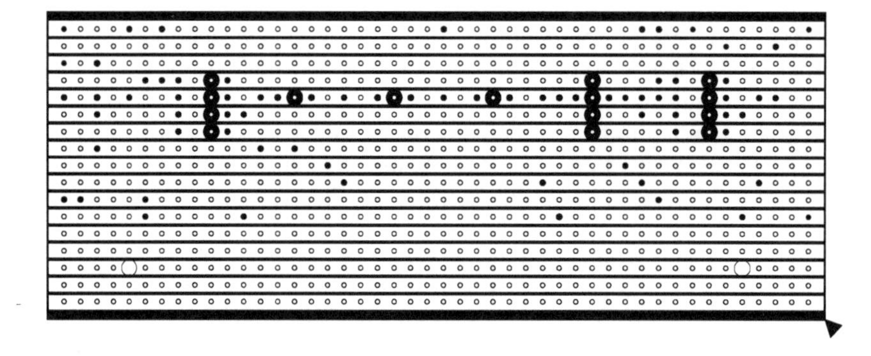

Figure 7.4 Underside view of the board

Construction and use

Refer to Figure 7.3 for details of the stripboard component layout and the hard wiring. The underside view of the board is provided in Figure 7.4. A board having 46 holes by 17 copper strips is required. The notes on building and using the scratch filter apply equally to this project, and should be consulted before building this unit and testing it.

Components list

Resistors (0.25 watt 5% carbon film except where noted)

R1	2k2
R2,3	100k 1% (2 off)
R4,6	8k2 (2 off)
R5	18k
R7,8	120k 1% (2 off)

Capacitors

C1	100n ceramic
C2	100u 16V radial elect
C3	470n polyester, 5mm lead spacing
C4,5,6,7	220n polyester, 5mm lead spacing (4 off)
C8	10u 25V radial elect

Semiconductors

IC1,2	TL071CP Bifet op amp, or similar (2 off)

Miscellaneous

S1,2	S.P.S.T. min toggle (2 off)
B1	9 volt (PP3 size)
SK1,2	Phono socket (2 off)

Metal case, 0.1 inch pitch stripboard measuring 46 holes by 17 copper strips, 8-pin d.i.l. holder (2 off), battery connector, wire, solder, etc.

8 RIAA preamplifier

Vinyl discs have now been superseded by compact discs, but with countless millions of vinyl records in existence they will remain in use for some years yet. A good magnetic cartridge represents a relatively cheap means of obtaining high quality results from gramophone records. A magnetic cartridge provides a typical output level of about 5 millivolts r.m.s., and requires a load impedance that is usually 50k or 100k. On the face of it there is no problem in producing a preamplifier having a suitable voltage gain and input impedance, but in practice there is a slight problem. In order to provide a flat frequency response a magnetic cartridge preamplifier must provide equalisation. This is needed in order to counteract the equalisation use during the recording process.

This RIAA equalisation consists of bass cut which is used to ensure that excessive groove modulation does not occur when the signal contains strong low frequency signals. The playback system must provide complementary bass boost so that a flat overall frequency response is obtained. High frequency boost is used during the recording process as part of a simple form of noise reduction. This boost is known as pre-emphasis, and it is balanced by high frequency cut (de-emphasis) during playback. This treble cut helps to reduce background 'hiss' caused by 'surface noise', or electrical noise in the early stages of the playback amplifier. Again, the playback response complements the recording response, giving a flat response overall.

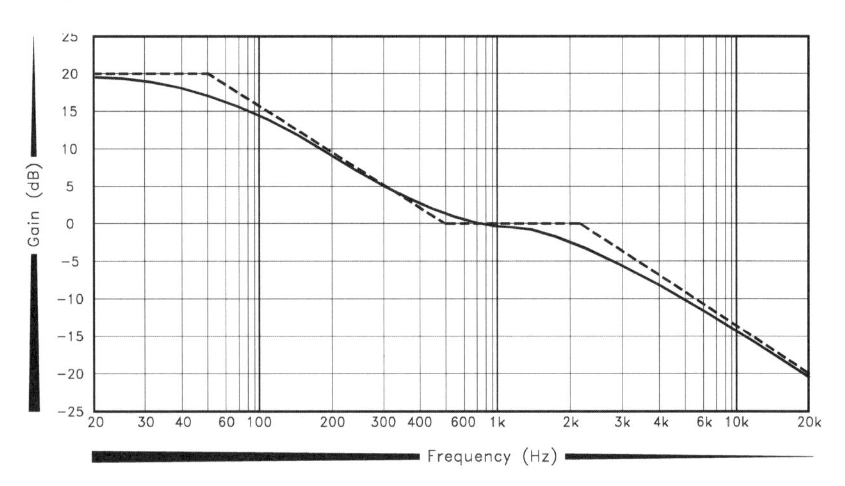

Figure 8.1 RIAA playback response

Figure 8.1 shows the standard RIAA playback response. The broken line shows the standard 'text book' version of the response, but in reality the filtering is not switched on and off at certain frequencies. The response changes in a more gradual fashion, as in the response indicated by the 'solid' line. The bass boost commences at 505Hz, and the treble cut starts at 2.12kHz. The response must be rolled-off at very low frequencies to avoid problems with infra-audio signals caused by warped records. This low frequency roll-off tends to occur naturally anyway, due to the affects of coupling capacitors in the circuit.

The circuit

The circuit diagram for a stereo RIAA playback preamplifier appears in Figure 8.2. We will only consider operation of the left hand channel, but the two channels are identical and operate in the same manner. IC1 is used as a low noise input stage which operates in the non-inverting mode and provides a voltage gain of 11. R1 shunts the basic 500k input impedance of the amplifier to provide an approximate input impedance of

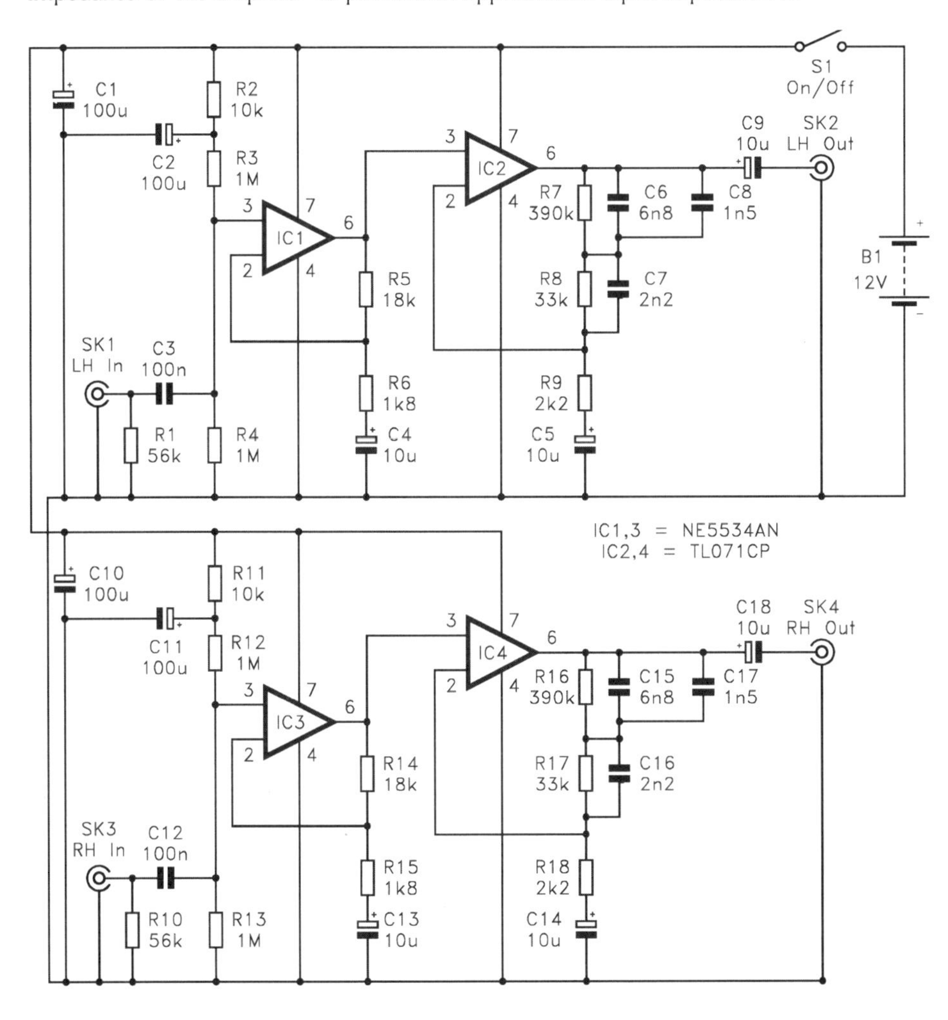

Figure 8.2 RIAA playback preamplifier circuit

50k. The value of R1 (and R10 in the right hand channel) should be increased to 120k if an input impedance of 100k is preferred.

A second non-inverting amplifier based on IC2 provides further amplification and implements the RIAA equalisation. At low audio frequencies the impedance through C6 to C8 is low in comparison to the impedance of R7 and R8, and the voltage gain of the circuit is therefore controlled by R7, R8, and R9. As the input frequency is increased, the impedance through C6 and C8 becomes low relative to the value of R7, giving increased feedback and reduced voltage gain. This gives the required reduction in gain from 50Hz to 505Hz. At high audio frequencies the impedance of C7 becomes low in comparison to the resistance of R8, again giving increased feedback and reduced voltage gain. This provides the reduction in gain from 2.12kHz to 20kHz.

The ideal time constant for R7 and C6/C8 is 3.18 milliseconds. Replacing the parallel capacitance of C6 and C8 (8n3) with an 8n2 capacitor gives slightly improved accuracy, but a capacitor of this value might be difficult to obtain. The ideal time constant for R8 and C7 is 75 microseconds. The specified values give a time constant of 72.6 microseconds, which is close enough to give good results. For the feedback network to give the desired result the value of R7 needs to be about 12 times higher than that of R8. In this case it is about 11.8 times higher, which is near enough.

The circuit has a current consumption of about 11 milliamps from a 12 volt supply. It will run from a nine volt battery, but a 12 volt type gives somewhat better performance, particularly in terms of overload margin. In fact there is some advantage in using a higher supply voltage of around 15 to 24 volts, but the electrolytic capacitors must have suitably high voltage ratings. It is not essential to use a battery, and a 12 volt mains power supply unit should give good results provided it has a reasonably well smoothed output.

Construction

The RIAA preamplifier is constructed on a stripboard which measures 38 holes by 39 copper strips. The component layout and wiring are shown in Figure 8.3, and the underside view of the board appears in Figure 8.4. Construction of the board follows along the usual lines. It might be worthwhile insulating the two long link-wires up the right hand side of the board, to make sure that they do not short-circuit to each other or any nearby components.

This preamplifier has a moderately high input impedance together with very high gain at low frequencies. This makes it very vulnerable to stray pickup of mains 'hum.' It is essential to house the unit in a good quality metal case that will provide good screening. In use the unit simply connects between the record deck and the amplifier, mixer, or whatever using good quality screened phono leads.

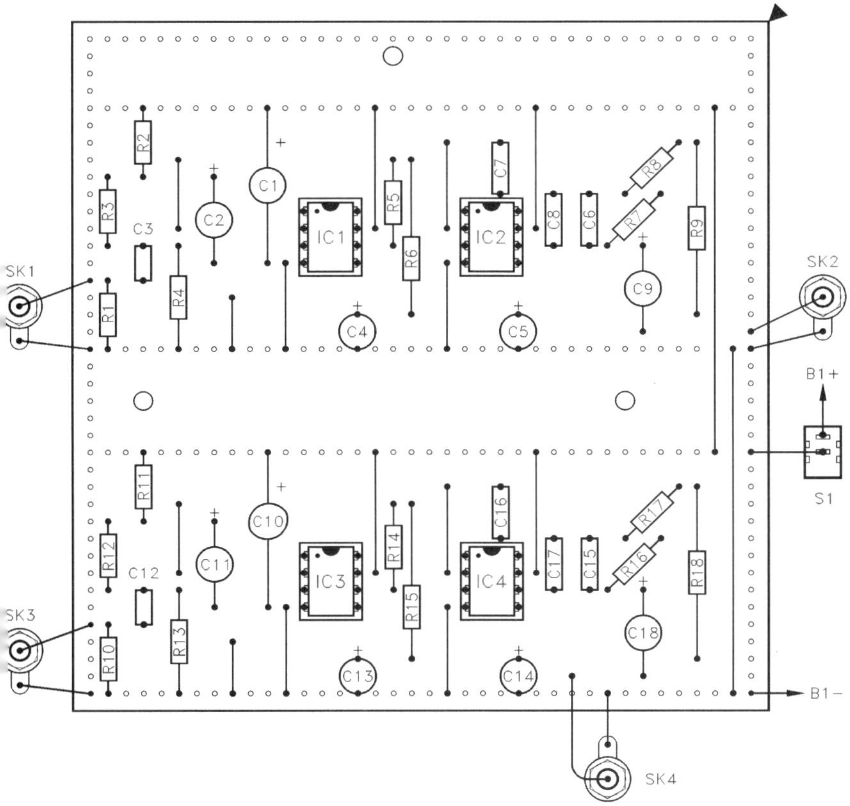

Figure 8.3 Component layout and wiring

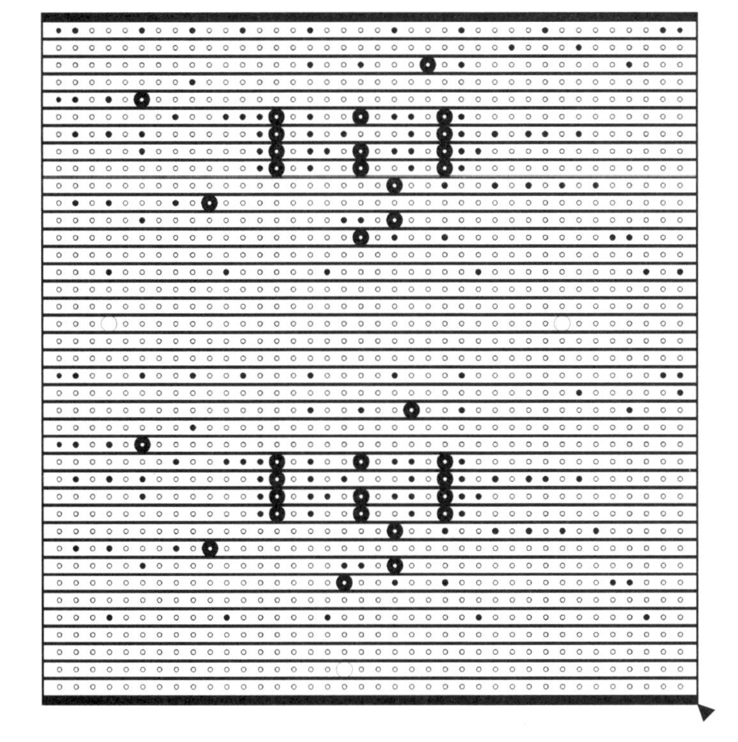

Figure 8.4 Underside view of the board. It measures 38 holes by 39 strips.

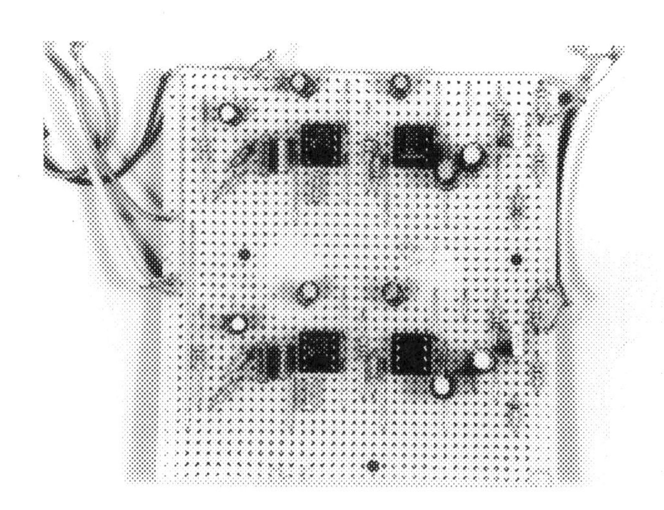

Photo 8.1 The completed RIAA preamp

Components list

Resistors (all 0.25 watt 5% carbon film)

R1,10	56k (2 off)
R2,11	10k (2 off)
R3,4,12,13	1M (4 off)
R5,14	18k (2 off)
R6,15	1k8 (2 off)
R7,16	390k (2 off)
R8,17	33k (2 off)
R9,18	2k2 (2 off)

Capacitors

C1,2,10,11	100u 16V radial elect (4 off)
C3,12	100n polyester, 5mm lead spacing (2 off)
C4,5,9,13,14,18	10u 25V radial elect (6 off)
C6,15	6n8 polyester, 7.5mm lead spacing (2 off)
C7,16	2n2 polyester, 7.5mm lead spacing (2 off)
C8,17	1n5 polyester, 7.5mm lead spacing (2 off)

Semiconductors

IC1,3	NE5534AN low noise op amp (2 off)
IC2,4	TL071CP Bifet op amp (2 off)

Miscellaneous

S1	S.P.S.T. min toggle switch
B1	12 volt (e.g. 8 x HP7 size cells in holder)
SK1,2,3,4	Phono socket (4 off)

Metal case, 0.1 inch matrix stripboard measuring 38 holes by 39 copper strips, battery connector, 8-pin d.i.l. holder (4 off), wire, solder, etc.

9

Tape preamplifier

New cassette tape mechanisms seem to be unobtainable at present, but 'surplus' mechanisms are often available at very low prices. Also, perfectly good and little used cassette mechanisms can sometimes be salvaged from a stereo cassette deck, radio cassette unit, etc. that has been dropped, or otherwise met a premature end. Producing your own recorder from a low cost or salvaged mechanism is perfectly possible, but ideally requires a fair amount of test equipment and the ability to use it.

On the other hand, a cassette player is easily constructed, and needs no adjustment once completed. However, before undertaking this project make sure that the cassette mechanism is in good working order, and that you can actually get it running properly with a suitable supply for the motor. Once the mechanical side of things is sorted out, and the motor is fed from a suitable supply, a simple preamplifier is all that is needed in order to produce a cassette player that can feed into a hi-fi amplifier.

The output level from a cassette playback head is very low, at under one millivolt r.m.s. for a mono head, and only half as much per channel for a stereo head. The preamplifier must therefore provide a large amount of voltage gain, but it does not have to provide a particularly high input impedance. An input impedance of a few kilohms gives good results. Matters are complicated slightly by the need for equalisation. In theory, the output level from a tape head rises at 6dB per octave (i.e. a doubling in the signal frequency gives a doubling of the output level). Also, pre-emphasis is used during the recording process. On the face of it, a massive amount of lowpass filtering is required during playback in order to give a flat frequency response overall.

In reality things are not as clear cut as this, because 'real world' tape heads do not achieve anything approaching theoretical perfection. This results in greatly reduced output at high frequencies, and the equalisation characteristic of the playback preamplifier has to be modified to allow for this.

The circuit

The circuit diagram for the tape preamplifier appears in Figure 9.1. This has obvious similarities to the RIAA preamplifier described previously, and it uses the same arrangement of a non-inverting input stage, followed by a second non-inverting amplifier which provides the equalisation. If we consider the circuit for the left hand channel, IC1 operates at a closed loop voltage gain of about 22, and provides an input impedance of just under 4.3k.

IC2 has a closed loop voltage gain of 561 at low frequencies, but C6 and R9 reduce the gain at higher frequencies. This gives the initial 6dB per octave roll-off to counteract the 6dB per octave increase from the tape head. At high frequencies R9 prevents any further roll-off and flattens the response. This allows for the fact that the output from the tape ceases to rise over this frequency range. In fact it reduces somewhat over the highest part of the audio range, but the pre-emphasis applied during the recording process counteracts this roll-off. This gives a reasonably flat

Figure 9.1 Circuit diagram for the tape preamplifier

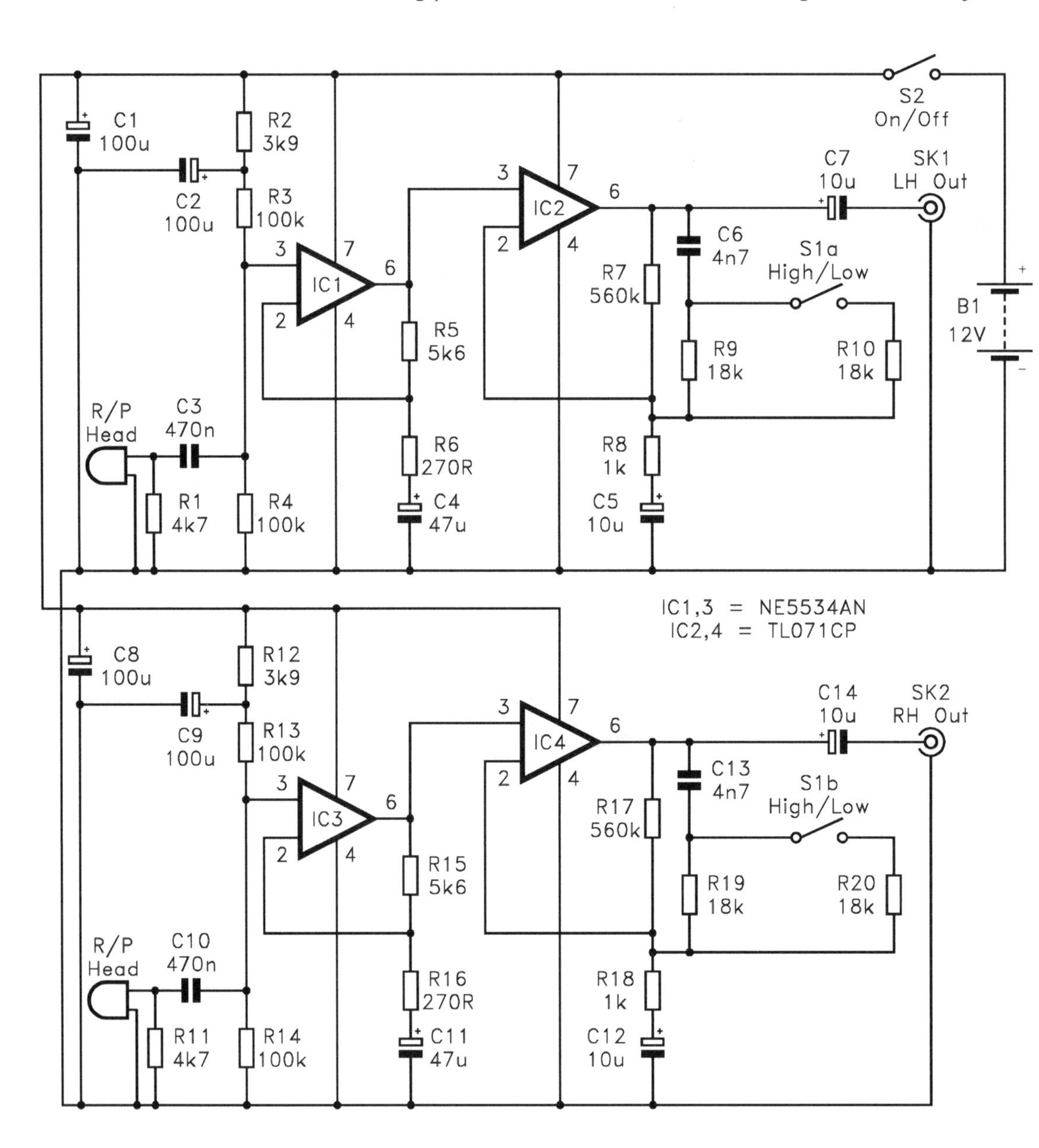

response, but the amount of treble produced is very much dependent on the quality of the tape and the tape head.

Most pre-recorded cassettes and home produced recordings are encoded with the Dolby B method of noise reduction. This means that some treble boost is applied to the recording at low to middle signal levels. The lower the signal level, the greater the amount of high frequency boost. Dolby B decoding is clearly desirable, but goes beyond the scope of a simple playback preamplifier. Results are quite good when Dolby B encoded tapes are played-back on equipment that does not include a suitable decoder, but overall there is a slight excess of treble. This can be counteracted by slightly more treble cut from the equalisation circuit, which gives a slightly flatter overall frequency response, and also gives a small amount of noise reduction. Closing S1 shunts R10 across R9, and gives a suitable equalisation characteristic.

The current consumption of the circuit is about 11 milliamps from a 12 volt supply. The circuit is shown as being battery powered, but it can be powered from a 12 volt mains power supply unit if preferred, but the power supply must provide a well smoothed output. If the cassette mechanism has a 12 volt motor it is possible to power the preamplifier and the motor from the same battery or mains power supply unit. However, electrical noise from the motor tends to find its way into the preamplifier via the supply lines. Making C1 and C8 about 10 times higher in value might reduce this problem to an insignificant level, but combating interference from the motor can be very difficult.

Steering around the problem by powering the preamplifier and the motor from separate batteries is probably the best way of obtaining a really low background noise level. The same approach can be used if the circuit is powered from a mains power supply unit. The two supply circuits can be powered from a common mains transformer provided they are fed from separate secondary windings.

Construction

Figure 9.2 and 9.3 respectively show the component and underside views of the stripboard, which measures 37 holes by 39 copper strips. Construction of this board is very straightforward and should not present any major difficulties. The point-to-point wiring is shown in Figure 9.4, which should be used in conjunction with Figure 9.2 (e.g. point 'A' in Figure 9.2 connects to point 'A' in Figure 9.4).

It is likely that the cassette mechanism will be fitted with a combined record/playback head, but in this case it is obviously used only as a playback head. The connections for the record/playback head are correct for the ones I have used, but it is possible that some heads have a different pin-out configuration. A little experimentation should soon sort things out if this method of connections fails to give good results. Do not be tempted to make continuity checks on the tape head. This could magnetise the head and impair its performance. Each channel of the preamplifier should be connected to the tape head via a separate screened lead. In order to

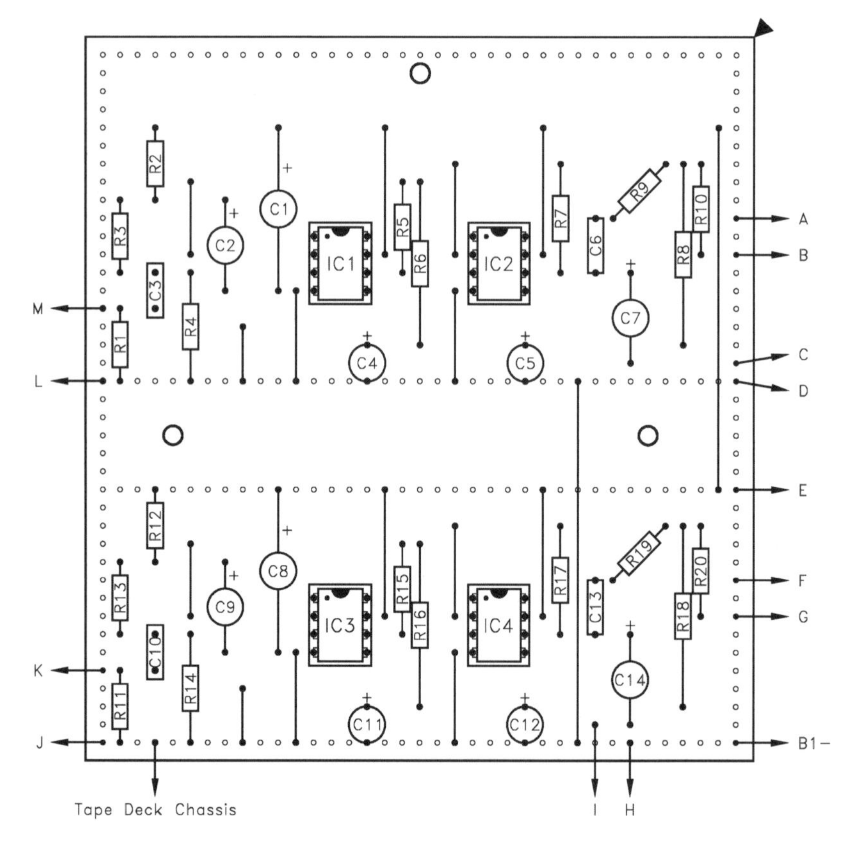

Figure 9.2 Component view of the stripboard

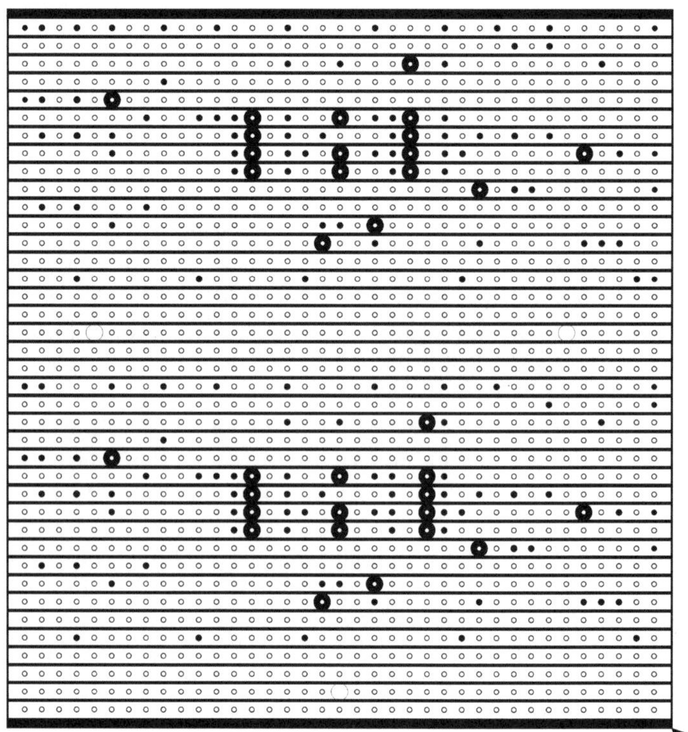

Figure 9.3 Underside view of the stripboard

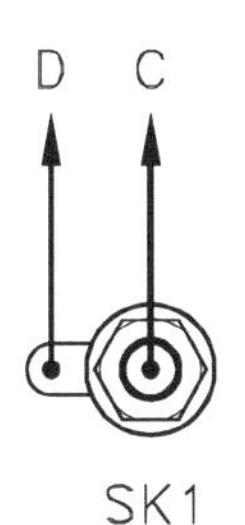

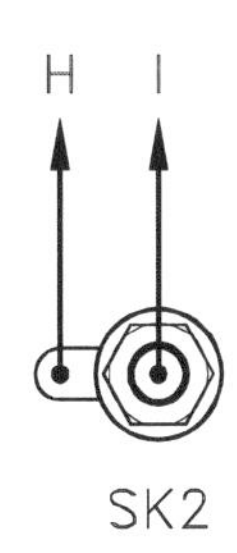

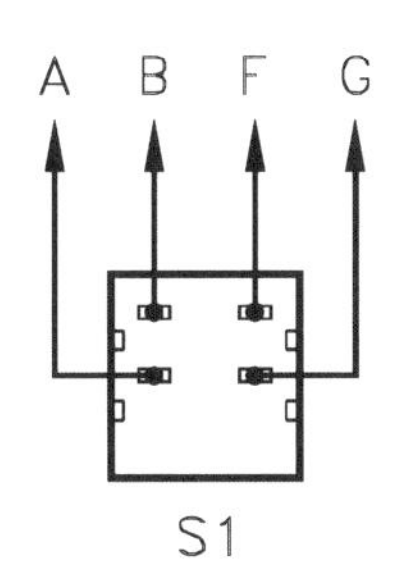

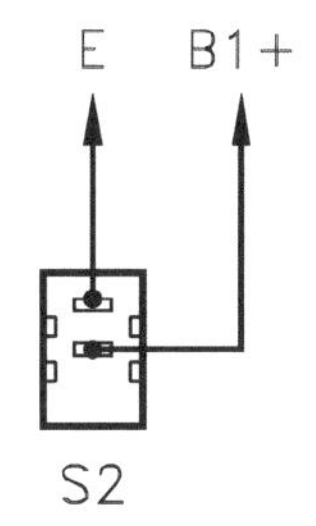

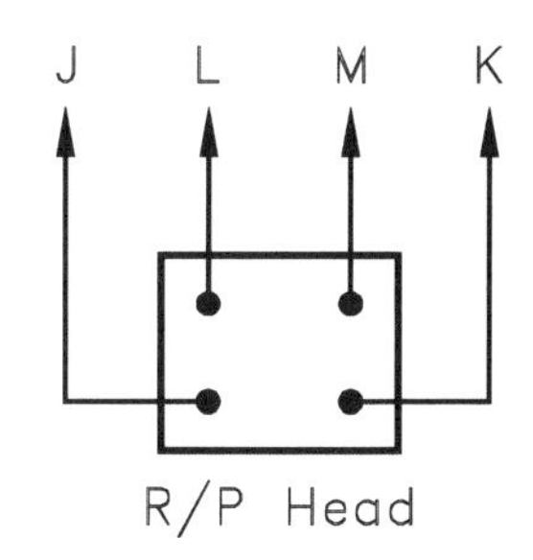

Figure 9.4 Point-to-point wiring

obtain a low noise level with insignificant 'humming' and 'buzzing' in the background, it will almost certainly be necessary to connect the 0 volt rail of the preamplifier to the chassis of the cassette mechanism. A suitable connection point is indicated in Figure 9.2.

Tweaking

Good results should be obtained with the specified values for R9/19 and C6/13, but it may well be possible to obtain improved results by 'tweaking' the values of these components. There is no standard playback response for a cassette tape preamplifier, because the ideal response depends on the characteristics of the tape head used. Hence some experimentation with the equalisation circuit can optimise results for a particular playback head.

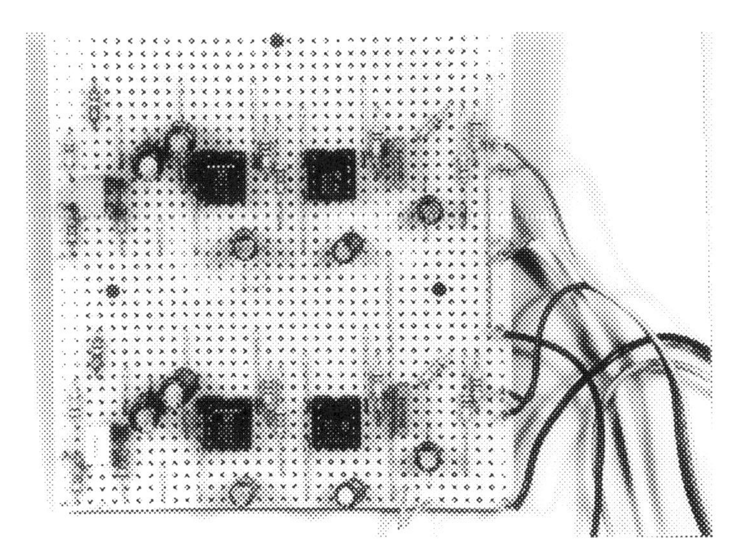

Photo 9.1 The finished tape preamplifier

Components list

Resistors (all 0.25 watt 5% carbon film)

R1,11	4k7 (2 off)
R2,12	3k9 (2 off)
R3,4,13,14	100k (4 off)
R5,15	5k6 (2 off)
R6,16	270R (2 off)
R7,17	560k (2 off)
R8,18	1k (2 off)
R9,10,19,20	18k (4 off)

Capacitors

C1,2,8,9	100u 16V radial elect (2 off)
C3,10	470n polyester, 5mm lead spacing (2 off)
C4,11	47u 25V radial elect (2 off)
C5,7,12,14	10u 25V radial elect (4 off)
C6,13	4n7 polyester, 7.5mm lead spacing (2 off)

Semiconductors

IC1,3	NE5534AN low noise op amp (2 off)
IC2,4	TL071CP Bifet op amp (2 off)

Miscellaneous

S1	D.P.S.T. min toggle switch
S2	S.P.S.T. min toggle switch
B1	12 volt (e.g. 8 x HP7 size cells in holder)
SK1,2	Phono socket (2 off)

Metal case, 0.1 inch matrix stripboard measuring 37 holes by 39 copper strips, battery connector, 8-pin d.i.l. holder (4 off), wire, solder, etc.

10

Audio limiter

An audio limiter is a form of automatic gain control. Under normal circumstances it does not actually do anything, and the input signal simply passes through to the output unprocessed. However, if the amplitude of the input signal exceeds a certain threshold level, the gain of the limiter reduces. The higher the input level rises above the threshold level, the greater the losses through the limiter. Once the threshold level has been reached, even large increases in the signal level fail to produce any significant increase in the output level. Once the input level has fallen back below the threshold level, the gain has risen to unity again, and the limiter allows the input signal to pass straight through to the output once more.

The limiter thus prevents the output signal from exceeding a certain level. The point of this is to prevent overloading of an amplifier, tape recorder, or some other device fed from the output of the limiter, if the input level should become higher than expected. Overloading causes clipping and severe distortion, which is avoided by a fast acting limiter circuit.

System operation

The block diagram of Figure 10.1 helps to explain the basic way in which the audio limiter functions. The main signal path is via a buffer amplifier, a VCA (voltage controlled attenuator), and a second buffer amplifier. The two buffer amplifiers simply ensure that the VCA is fed from a suitably low source impedance, and that it drives a high impedance load. This gives minimal losses under normal operating conditions, with the circuit effectively operating as a buffer stage. The VCA provides low losses with zero control voltage, but provides increasing attenuation as the control voltage is increased.

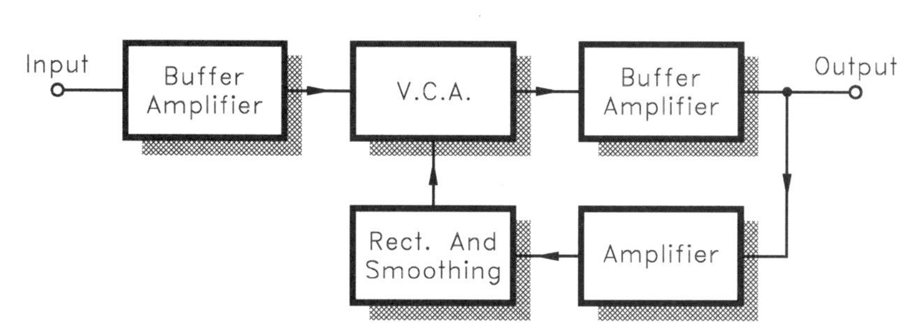

Figure 10.1 Block diagram of the audio limiter

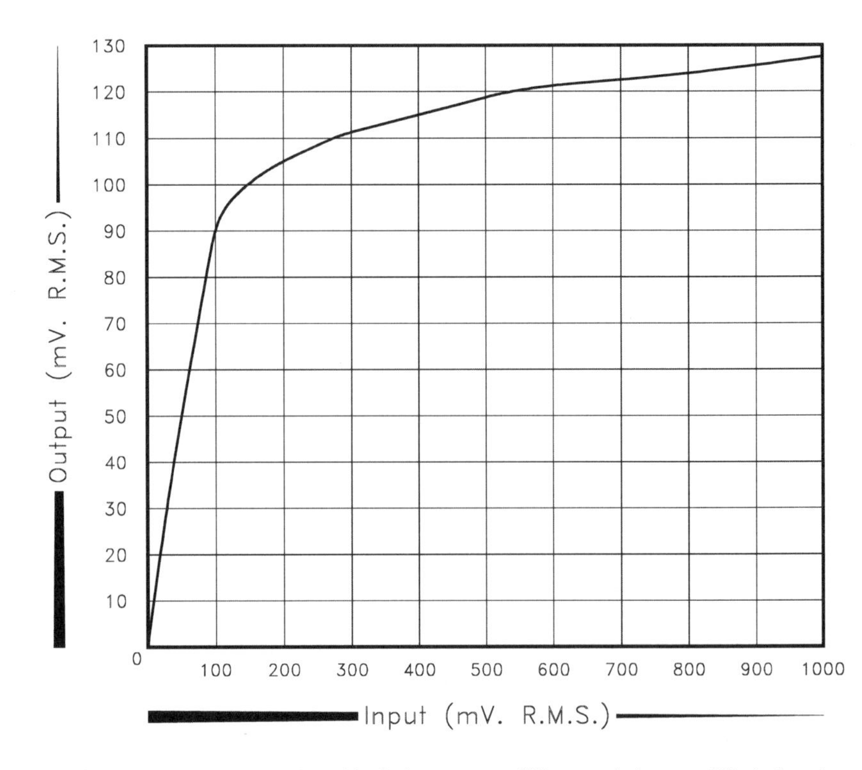

Figure 10.2 Compression characteristic of the prototype limiter

Some of the output signal is fed to an amplifier, and the amplified signal is rectified and smoothed. This produces a positive output voltage that is roughly proportional to the amplitude of the input signal. There is not a precise linear relationship between the amplitude of the input signal and the output potential from the smoothing circuit due to small but significant voltage drops through the diodes in the rectifier circuit. This means that low level input signals fail to produce any output from the smoothing circuit. The control characteristic of the VCA is such that small input voltages have no affect. Consequently, the output signal has to reach a fairly high level before the control voltage to the VCA produces an increase in the attenuation level. However, once this point is reached, any rise in the output level is counteracted by an increase in the attenuation through the VCA.

The output level does still increase, but it lags far behind the input level. Figure 10.2 shows the compression characteristic obtained from the prototype limiter (using a 1kHz sinewave test signal). At input levels of up to about 90 millivolts there is approximately unity voltage gain through the circuit, but at higher input levels the compression comes into operation. Increasing the input level by a factor of more than 10 gives an increase in the output level of under 50 percent.

VCA

The task of producing a simple low distortion VCA may seem to be simple enough using modern semiconductor circuits, but it is something that is problematic. It is easy enough to use a Jfet or MOSFET as a voltage con-

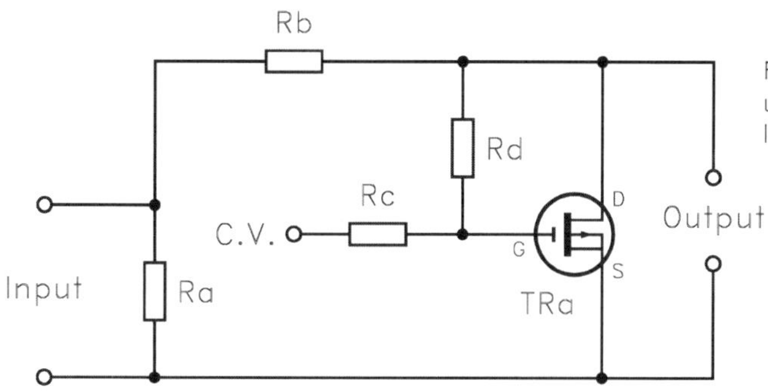

Figure 10.3 A simple VCA using feedback to improve linearity

trolled resistor in a VCA circuit, but with a basic circuit there are problems with distortion. The main problem occurs when the Jfet or MOSFET starts to turn on. The negative half cycles become clipped, and although the clipping is of the 'soft' variety, it produces quite strong distortion products. The distortion is often around five to 10 percent, and can be even higher. This is clearly audible on most types of signal.

It is possible to reduce the distortion by using feedback from the drain to the gate of the f.e.t., as shown in the VCA circuit of Figure 10.3. The attenuator is formed by Rb and the drain-to-source resistance of TRa. In this example TRa is an N channel MOSFET, and it is therefore switched off with its gate at 0 volts. This gives minimal losses through Rb. If the gate of TRa is steadily taken more positive, the device will turn on, and its drain-to-source resistance will eventually fall to just a few hundred ohms. Provided the value of Rb is at least a few tens of kilohms, this gives losses of about 40dB or more through Rb (i.e. reduces the signal voltage by a factor 100 or more).

The feedback from the drain to the gate is provided by Rc and Rd. For Jfets it is normally recommended that Rc should have the same value as Rd, but in practice this often overcompensates and moves the 'soft' clipping from the negative half cycles to the positive ones. For MOSFETs this ratio usually seems to be inadequate, with the 'soft' clipping of the negative half cycles being lessened but still pretty obvious. Making Rc about three to four times higher in value than Rd seems to give much better results, with low distortion. One problem in adding the feedback is that it couples the control voltage into the audio signal path. However, the degree of coupling is kept very low provided Rc and Rd are much higher in value than Ra and Rb.

The circuit

Refer to Figure 10.4 for the full circuit diagram of the audio limiter. This circuit is for a mono unit, but stereo limiters have totally separate processing for the two channels. Therefore, for stereo operation it is merely necessary to build two limiter circuits, one for each stereo channel.

IC1 operates as the input buffer stage, and it provides an input impedance of 50k. The VCA is based on IC2, and it uses the configuration

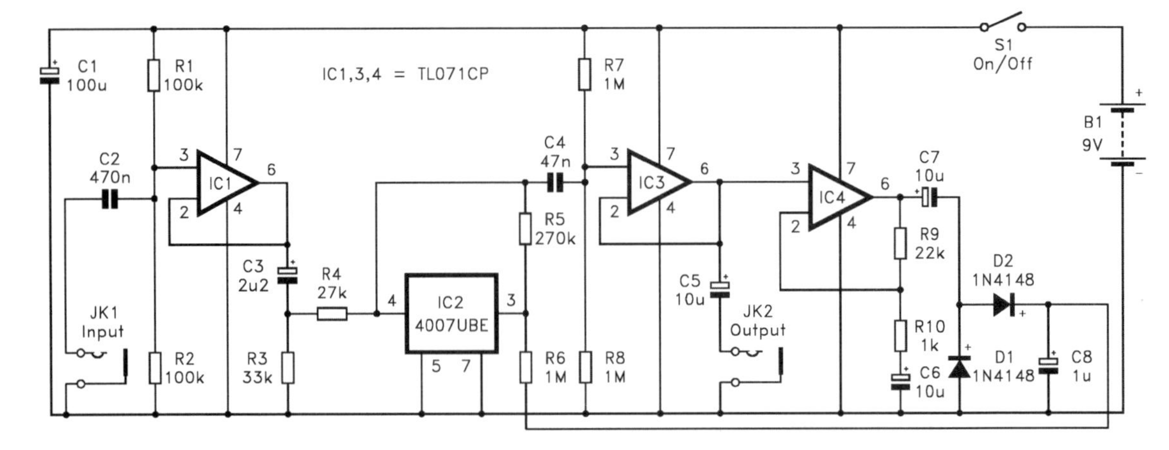

Figure 10.4 Full circuit diagram of the audio limiter

shown in Figure 10.3 and described previously. IC2 is actually a CMOS dual inverter and complementary pair, but in this case it is used as a simple N channel MOSFET. The drain, gate, and source terminals are at pins 4, 3, and 5/7 respectively. IC3 is used in the output buffer stage. This provides an input impedance of 500k, which ensures that only very low losses are introduced by R4 when IC2 is switched off. With IC2 biased into saturation losses of about 30 to 40 dB are produced through R4, which is more than adequate for this application.

The output of IC3 is amplified by IC4, which operates as a non-inverting amplifier having a closed loop voltage gain 23. The voltage gain controls the level at which the limiting is applied, and it is therefore possible to alter this level to some degree by altering the value of R9. Reducing its value increases the limiting threshold, and increasing its value reduces the threshold. C7 couples the output of IC4 to a simple half-wave rectifier and smoothing circuit based on D1 and D2. The low output impedance of IC4 gives a fast attack time so that the circuit responds almost instantly to high input levels. In order to avoid distortion it is essential to use a much longer decay time. The specified value for C8 gives a decay time of around one second, which will give good results in most applications. However, the decay time is easily altered, and is proportional to the value of C8 (e.g. a value of 0u47 gives a decay time of just under half a second).

The current consumption of the circuit is about 4.5 milliamps from a 12 volt supply. A mains power supply unit can be used to power the unit, but the supply must provide a well smoothed output.

Construction and use

Figure 10.5 shows the component layout and hard wiring for the audio limiter. The underside view of the board appears in Figure 10.6. This layout is based on a board which measures 52 holes by 18 copper strips. In most respects construction of the board is very straightforward, but bear in mind that the 4007UBE used for IC2 is a CMOS device, and that it requires the standard anti-static handling precautions. On the prototype

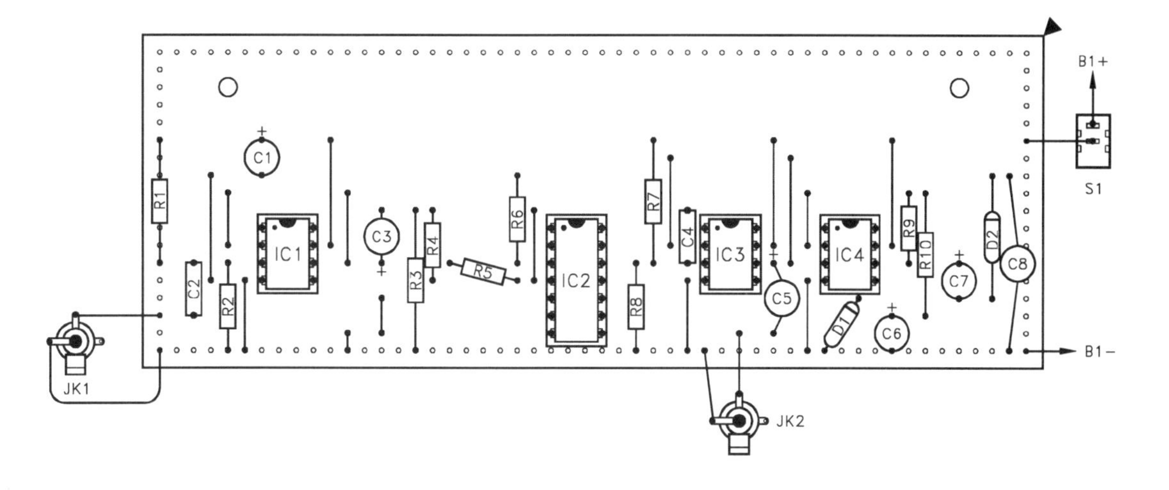

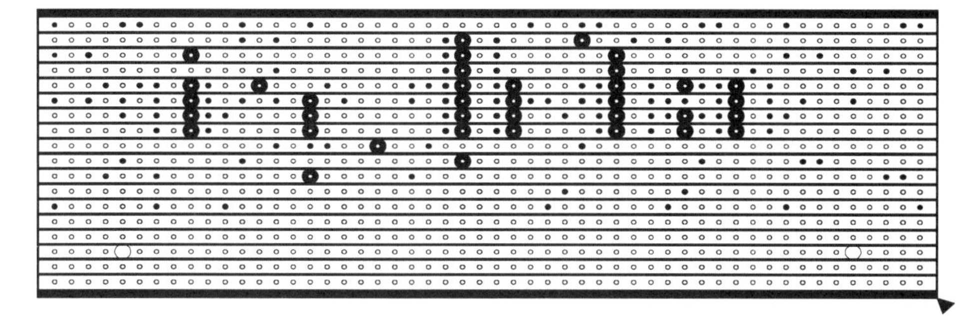

JK1 and JK2 are 3.5 millimetre jack sockets, but these can obviously be changed to any audio connector that fits in well with your equipment.

To test the unit feed the input with a signal at a fairly high level of around 500mV to one volt r.m.s., and monitor the output. This should give a strong limiting effect, and the background noise level will probably be heard to rise during periods when the signal level is relatively low. It is helpful here to use a signal that has a relatively low signal to noise ratio, such as an f.m. tuner tuned to a weak station, or an old cassette recording. Temporarily short circuiting C8 will remove the compression, and should result in a large increase in the output level.

Top – Figure 10.5 Component layout and hard wiring of the audio limiter
Lower – Figure 10.6 Underside view of the board of the audio limiter. This has 52 holes by 18 copper strips

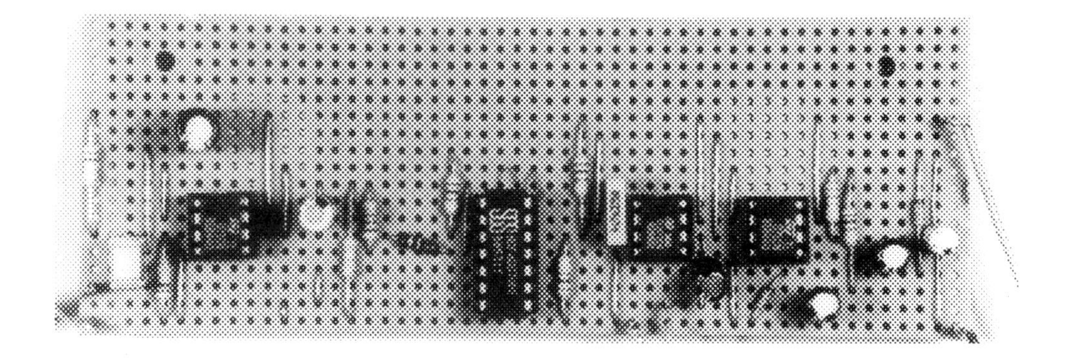

The finished audio limiter

Components list

Resistors (all 0.25 watt 5% carbon film)

R1,2	100k (2 off)
R3	33k
R4	27k
R5	270k
R6,7,8	1M (3 off)
R9	22k (see text)
R10	1k

Capacitors

C1	100u 16V radial elect
C2	470n polyester, 7.5mm lead spacing
C3	2u2 50V radial elect
C4	47n polyester, 7.5mm lead spacing
C5,6,7	10u 25V radial elect (3 off)
C8	1u 50V radial elect

Semiconductors

IC1,3,4	TL071CP Bifet op amp (3 off)
IC2	4007UBE CMOS dual inverter and comp. pair
D1,2	1N4148 silicon diode (2 off)

Miscellaneous

JK1,2	3.5mm jack socket (2 off)
B1	12 volt (e.g. 8 x HP7 size cells in holder)
S1	S.P.S.T. min toggle switch

Metal case, 0.1 inch matrix stripboard measuring 52 holes by 18 copper strips, battery connector, 8-pin d.i.l. holder (3 off), 14-pin d.i.l. holder, wire, solder, etc.

11

Bass and treble tone controls

This circuit is for conventional bass and treble tone controls which can provide boost or cut. Figure 11.1 shows the responses obtained with both tone controls set for maximum boost and for maximum cut. As will be apparent from this, up to about 15dB of boost and cut are available at the extremes of the audio range. With both controls centred, the frequency response is flat over the audio range, and the circuit provides unity voltage gain.

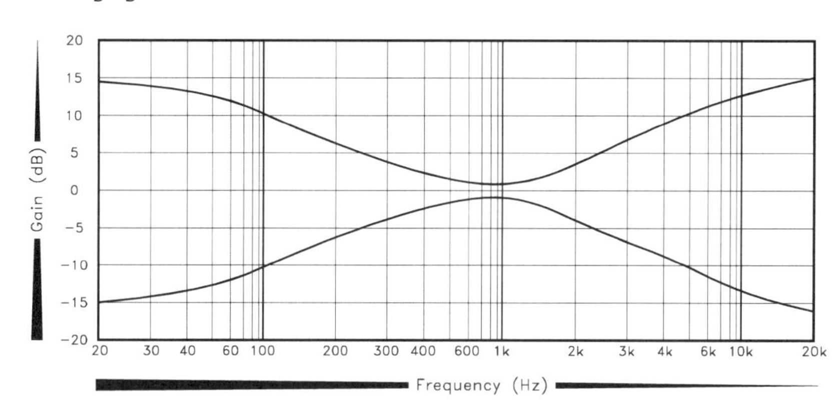

Figure 11.1 Responses obtained with both tone controls set for maximum boost and for maximum cut

The circuit

The circuit diagram for the bass and treble tone controls appears in Figure 11.2. IC1 merely acts as an input buffer stage which provides an input impedance of 50k, and a low enough output impedance to drive the tone control stage properly. The tone control circuit is an active type based on IC2. IC2 operates in the inverting mode, and it has the tone control network in its feedback circuit. Increased feedback through the tone control network gives reduced voltage gain, and reduced feedback gives increased voltage gain.

VR1 is the bass control, and VR2 is the treble control. If we take the bass control first, and ignore C5 and C6 for the moment, we have a simple negative feedback network that produces increased gain if the wiper of VR1 is taken towards the left hand end of the track, or decreased gain if it is taken towards the other end of the track. Due to the inclusion of C5 and C6 the circuit only functions this way at low frequencies. At higher

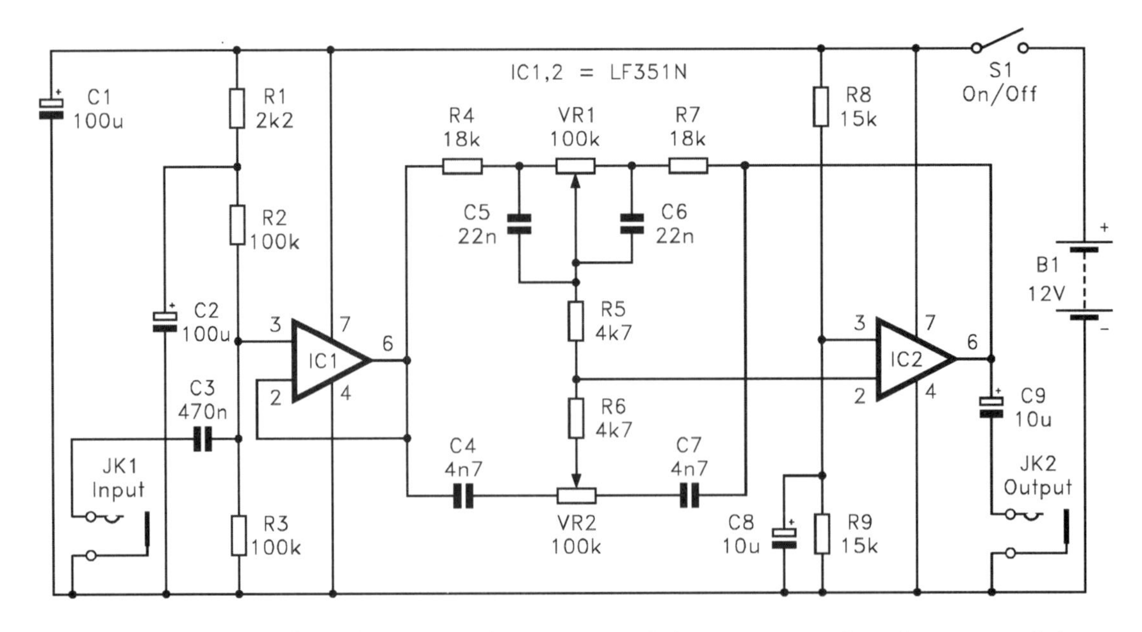

Figure 11.2 Circuit diagram for the bass and treble tone controls

frequencies the reactance of these two capacitors become much lower, and they effectively short circuit VR1. The voltage gain of the circuit is then set at unity by R4 and R7.

The treble circuit also relies on the fact that the reactance of a capacitor decreases as the applied frequency is increased. At high frequencies C4 and C7 have a reactance value that is low in comparison to the track resistance of VR2. The voltage gain of the circuit is therefore largely controlled by VR2. At low frequencies the reactance of C4 and C7 becomes relatively high. With these two components providing a reactance of several hundred kilohms of more, the setting of VR2 can provide little influence on the gain of the circuit. The voltage gain is therefore about unity at middle and low frequencies, and controlled by VR2 at high frequencies. R5 and R6 are needed to minimise interaction between the two tone control networks.

A supply voltage of nine to 15 volts is required, although the circuit will operate with a supply potential of up to 30 volts if C1 and C2 have suitably high voltage ratings. The current consumption is only about four milliamps from a 12 volt supply. A mono tone control circuit is featured here, but for stereo operation it is merely necessary to build two of these circuits, one for each stereo channel.

Construction

A stripboard measuring 47 holes by 20 copper strips is used for this project. The component layout is shown in Figure 11.3, and the copper side view of the board appears in Figure 11.4. The hard wiring is shown in Figure 11.5, which must be used in conjunction with Figure 11.3. Construction of the board is very straightforward, and should present no difficulties. JK1 and JK2 are 3.5 millimetre jack sockets on the prototype,

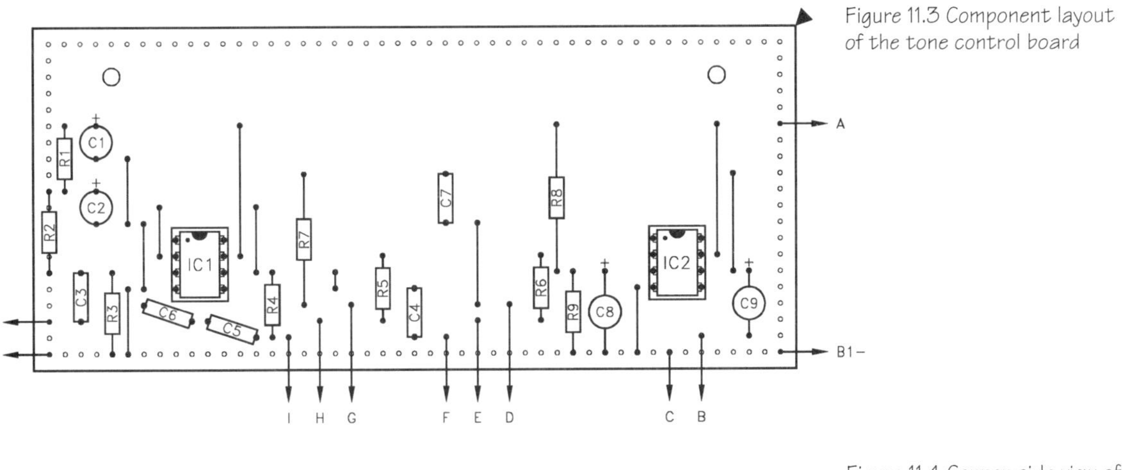

Figure 11.3 Component layout of the tone control board

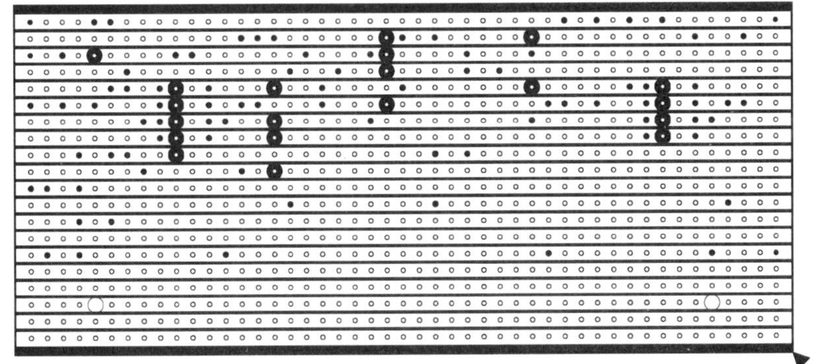

Figure 11.4 Copper side view of the board

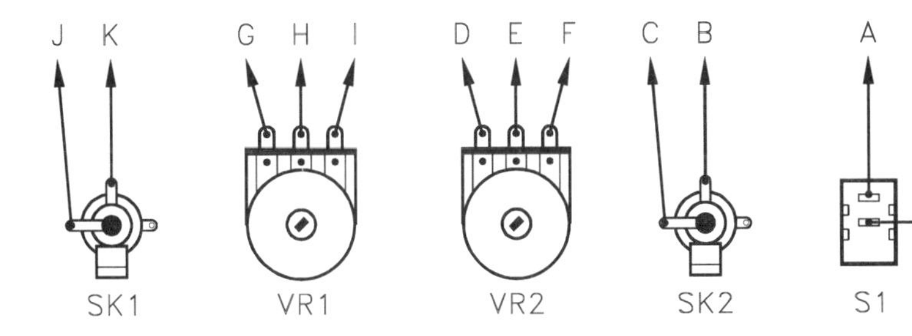

Figure 11.5 Hard wiring (use in conjunction with Figure 11.3)

but it is in order to use any preferred audio connectors instead. Obviously a circuit of this type will often form part of a more major piece of equipment, rather than being built as a stand-alone unit. The input and output sockets will then be unnecessary, and the circuit will be hard wired into the main unit.

Components

Resistors (all 0.25 watt 5% carbon film)

R1	2k2
R2,3	100k (2 off)
R4,7	18k (2 off)
R5,6	4k7 (2 off)
R8,9	15k (2 off)

Potentiometers

VR1,2	100k lin carbon rotary (2 off)

Capacitors

C1,2	100u 16V radial elect
C3	470n polyester, 7.5mm lead spacing (2 off)
C4,7	4n7 polyester, 7.5mm lead spacing (2 off)
C5,6	22n polyester, 7.5mm lead spacing (2 off)
C8,9	10u 25V radial elect (2 off)

Semiconductors

IC1,2	LF351N Bifet op amp (2 off)

Miscellaneous

S1	S.P.S.T. min toggle switch
B1	12 volt (e.g. 8 x HP7 size cells in holder)
JK1,2	3.5mm jack socket (2 off)

Metal case, 0.1 inch pitch stripboard measuring 47 holes by 20 copper strips, 8-pin d.i.l. holder (2 off), battery connector, control knob (2 off), wire, solder, etc.

12 Loudness filter

The term 'loudness filter' gives the impression that this type of circuit is either used when operating audio equipment at high volumes, or that it actually removes loud signals. In fact it is a simple filter which provides small amounts of bass and treble boost, and it is used when operating audio equipment at low volumes. It seems that the human hearing mechanism does not operate efficiently at low volumes, with very high and low frequency sounds tending to be missed. This is perhaps due to the fact that human hearing is relatively insensitive at the extremes of the audio range. Whatever the cause, a loudness filter boosts high and low frequencies to counteract the rather 'flat' sound that otherwise occurs with the volume set well back.

There is no standard characteristic for a loudness filter, but only a small amount of boost is needed in order to give the desired effect. Figure 12.1 shows the frequency response of the loudness filter featured here. The treble boost is applied at frequencies above and below about 800Hz, but the response has very gentle slopes. The maximum bass boost is just under 6dB, and the treble boost reaches a maximum of just 5dB at 20kHz. This very modest amount of filtering gives good results in practice.

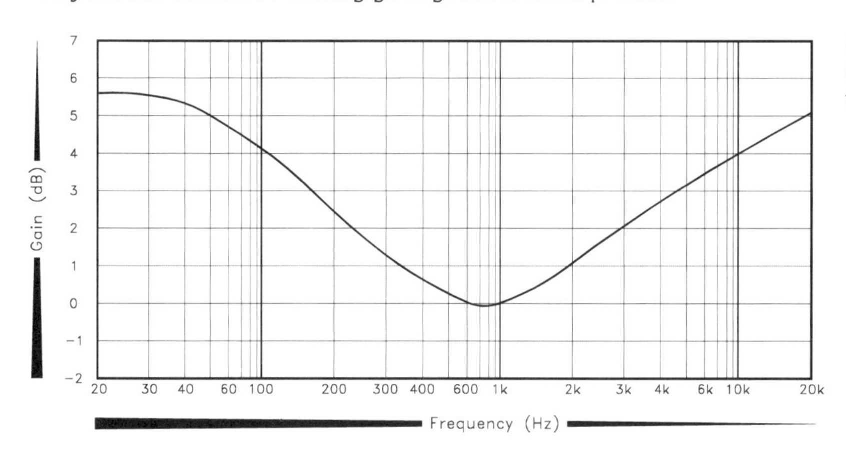

Figure 12.1 Frequency response of the loudness filter

The circuit

The loudness filter circuit diagram is provided in Figure 12.2. The filtering is provided by IC1, which operates as a non-inverting amplifier. R1 and R2 bias the non-inverting input of IC1 and set the input impedance of

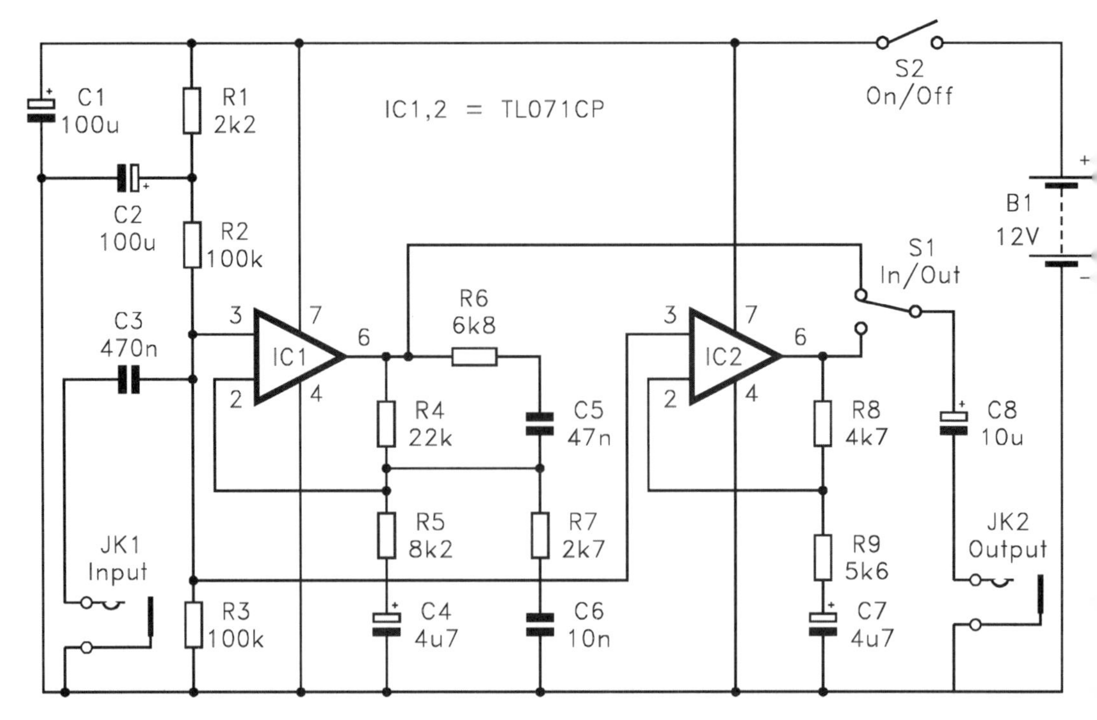

Figure 12.2 Loudness filter circuit diagram

the circuit at 50k. R4 and R5 set the closed loop voltage gain at about 3.7, but the gain is also affected by the frequency dependent feedback via R6, C5, R7, and C6. At high frequencies the reactance of C5 is very low, and R6 is effectively shunted across R5. This reduces the voltage gain of the circuit to a little under two times. The reactance of C5 becomes much higher at low frequencies, and this reduces the influence of R6. The gain of the circuit therefore increases at low frequencies.

At low frequencies C6 has a reactance that is high in comparison to the value of R5. R7 and C6 therefore have little influence on the voltage gain of IC1. At higher frequencies the reactance of C6 decreases, effectively shunting R7 across R5, and boosting the voltage gain of the circuit. This gives the required high frequency boost.

IC2 operates as a non-inverting amplifier having a voltage gain of just under two. It is biased from the same potential divider network as IC1, and it is also fed with the input signal. S1 enables either the filtered output from IC1 or the 'straight' signal from IC2 to be selected. This enables the loudness filtering to be switched in and out, with no significant change in volume when the switch is made.

The current consumption of the circuit is about four milliamps from a 12 volt supply. The circuit will operate with supply voltages from nine to 15 volts, or up to 30 volts if C1 and C2 have suitably high voltage ratings. A mains power supply unit can be used provided it has a reasonably well smoothed output.

Construction

A stripboard having 44 holes by 17 copper strips is used for the loudness filter. Figure 12.3 shows the component layout and wiring, while the underside view is provided in Figure 12.4. Construction of this board should be very straightforward, but as always, take care to fit the integrated circuits and electrolytic capacitors the right way round. For stereo operation two boards are required, one for each stereo channel.

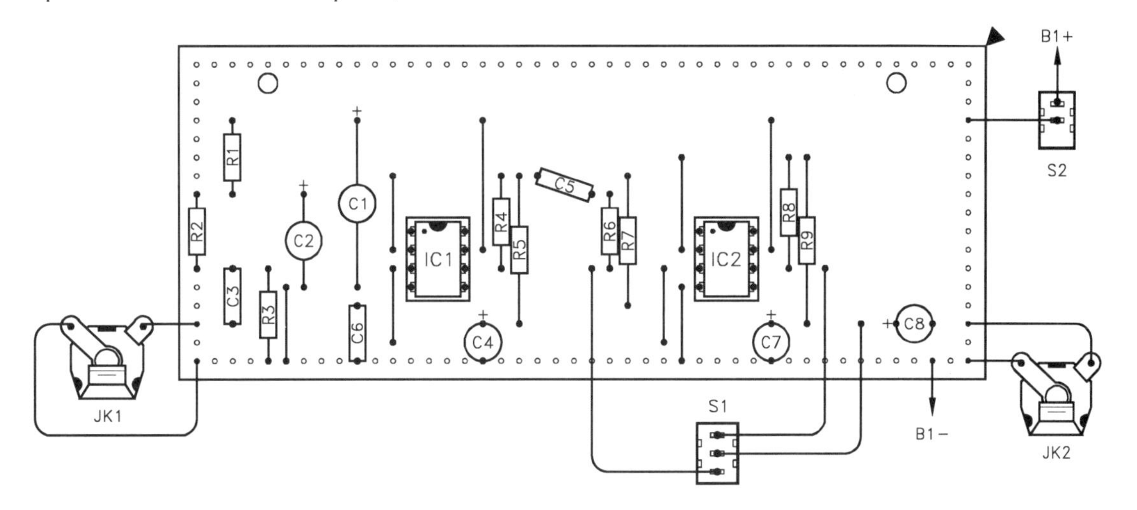

Figure 12.3 Component layout and wiring

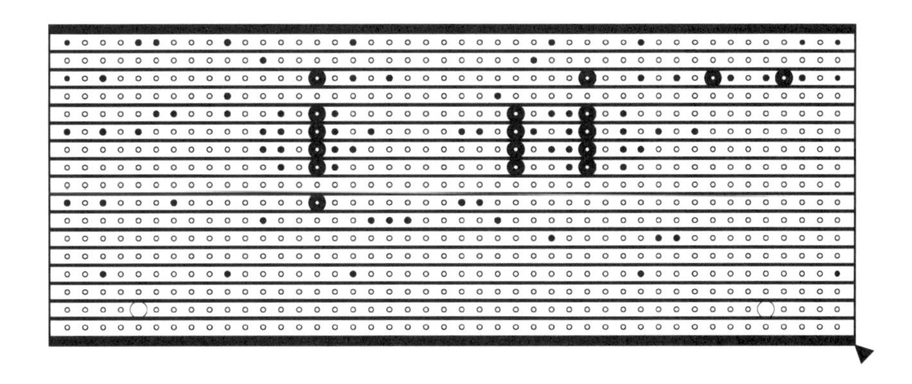

Figure 12.4 Underside view

JK1 and JK2 are standard jack sockets on the prototype, but any preferred type of audio connector can be used instead. Obviously a circuit of this type will often be part of a larger device such as an integrated preamplifier and power amplifier. JK1 and JK2 are then omitted, and the unit is hard wired into the main unit.

Components list

Resistors (all 0.25 watt 5% carbon film)

R1	2k2
R2,3	100k (2 off)
R4	22k
R5	8k2
R6	6k8
R7	2k7
R8	4k7
R9	5k6

Capacitors

C1,2	100u 16V radial elect (2 off)
C3	470n polyester, 7.5mm lead spacing
C4,7	4u7 50V radial elect (2 off)
C5	47n polyester, 7.5mm lead spacing
C6	10n polyester, 7.5mm lead spacing
C8	10u 25V radial elect

Semiconductors

IC1,2	TL071CP (2 off)

Miscellaneous

S1	S.P.D.T. min toggle switch
S2	S.P.S.T. min toggle switch
JK1,2	Standard jack socket (2 off)
B1	12V (e.g. 8 x HP7 size cells in holder)

Metal case, 0.1 inch pitch stripboard measuring 44 holes by 17 copper strips, 8-pin d.i.l. holder (2 off), battery connector, wire, solder, etc.

13

Loudness control

A loudness control is just a volume control that has built-in loudness filtering. In other words, as the volume is backed off, the bass and (or) treble is automatically boosted slightly. Adding automatic treble boost to a volume control is very simple, and merely requires the addition of one resistor and one capacitor, as shown in the circuit of Figure 13.1.

This type of control would be out of place in true hi-fi equipment, but can be useful addition to portable audio equipment that will often be used at relatively low volume levels.

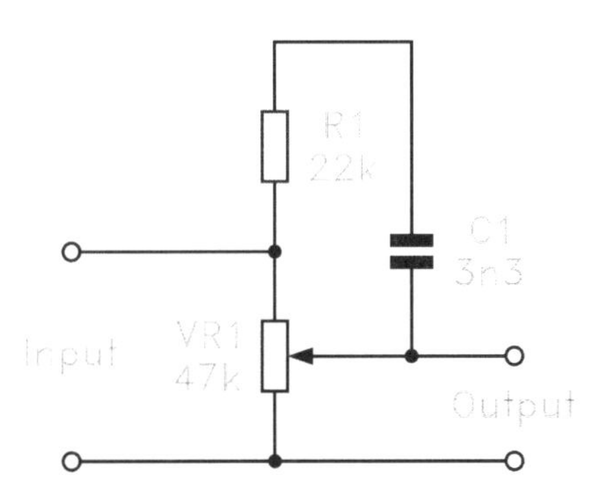

Figure 13.1 Adding simple loudness filtering to a volume control

VR1 is the existing volume control. At low and middle frequencies the reactance of C1 is high in comparison to the value of VR1, and the additional components have no significant affect. At higher frequencies the reactance of C1 becomes much lower, and it bypasses the upper section of VR1's track. This gives a certain amount of high frequency boost, but the amount of boost depends on the setting of VR1. If it is set for high volume, the resistance from its wiper to the upper track connection will be quite low. It is only at the highest audio frequencies that C1 will have a low enough reactance to give a significant amount of boost. Even then, the series resistance of R1 will prevent C1 from producing more than very mild high frequency boost. At low volume settings there is a much higher resistance from VR1's wiper to the top end of its track, and C1 can then produce more boost, and over a wider range of frequencies. However, R1 still tames the boost, which will generally be no more than about 6dB to 10dB at the highest audio frequencies.

The specified values for C1 and R1 are correct for a 47k volume control. This table provides the appropriate values for R1 and C1 for other volume control values.

Values of R1 and C1 for various volume control values

Vol. Control	*R1*	*C1*
4k7	2k2	33n
10k	4k7	15n
22k	12k	6n8
100k	47k	1n5
220k	120k	680p
470k	220k	330p
1M	470k	150p
2M2	1M2	68p

The additional resistor and capacitor are simply wired across the appropriate two tags of the volume control, as shown in Figure 13.2

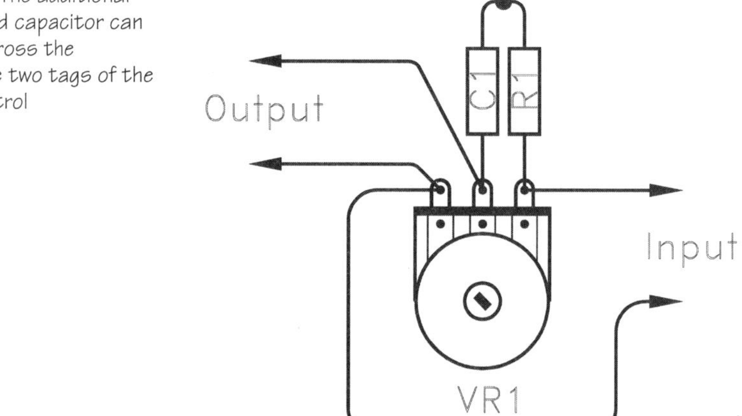

Figure 13.2 The additional resistor and capacitor can be wired across the appropriate two tags of the volume control

14

Simple graphic equaliser

A high performance graphic equaliser which covers the audio range in about eight to 10 bands is a highly desirable piece of equipment, but is also quite complex and expensive to build. Such a device is certainly beyond the scope of a book such as this. However, a relatively simple unit which covers the audio range in four or five bands is relatively simple and inexpensive, but still gives far more control over the sound than basic bass and treble tone controls. The unit featured here is a four band graphic equaliser which has approximate centre frequencies of 75Hz, 350Hz, 1.7kHz, and 7.5kHz. These correspond to the bass, lower middle frequency, upper middle frequency, and treble ranges. The maximum cut and boost available from each control is over 12dB. A flat frequency response is obtained with all four controls at a middle setting.

The circuit

Refer to Figure 14.1 for the circuit diagram of the four channel graphic equaliser. IC1 simply acts as an input buffer stage which provides an input impedance of 50k. This is followed by four filter stages that are essentially the same, but have different capacitor values so that they operate over different frequency bands.

If we consider the stage based on IC2a, this is a form of inverting mode amplifier. R4 and R7 set the voltage gain at unity, but frequency

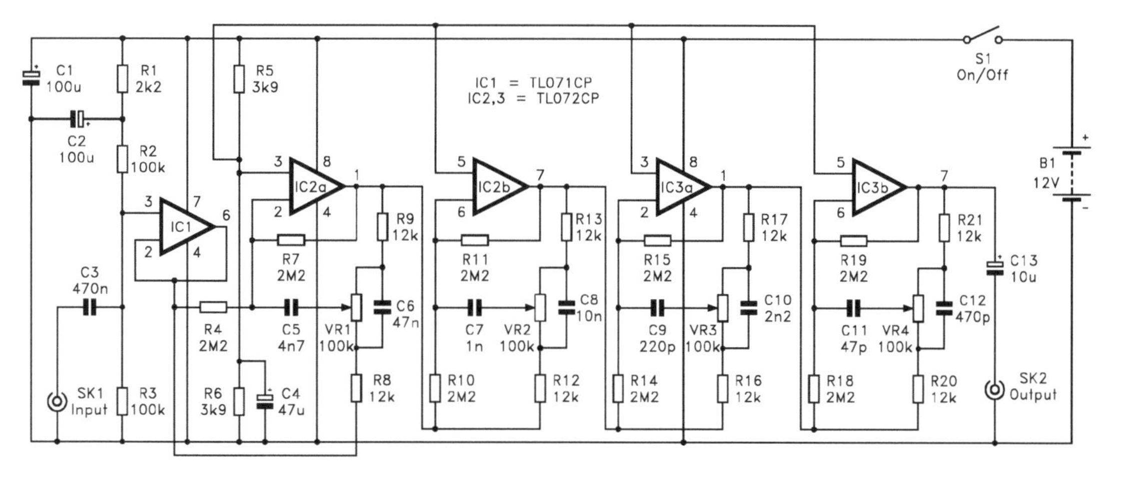

Figure 14.1 Circuit diagram of the four channel graphic equaliser

selective negative feedback is provided by C5, C6, R8, R9, and VR1. Operation of the circuit is easier to understand if C6 is ignored initially. With the wiper of VR1 at the top of its track, R9 and C5 are shunted across R7. At audio frequencies C5 has a very low reactance in comparison to the value of R7, and effectively shunts R9 straight across R7. This gives increased feedback, and greatly reduced voltage gain. If the wiper of VR1 is moved down the track, part of the track resistance is added in series with R9, and the drop in gain is reduced. With the wiper of VR1 at the middle of the track the feedback circuit is symmetrical, giving unity gain. Moving the wiper further down the track results in R8 and C5 being gradually shunted across R4. This gives reduced feedback and increased voltage gain.

As described so far, the circuit controls the voltage gain over a wide range of frequencies. In fact the gain is controlled at all frequencies above a certain figure. We require the gain control to be operative over a relatively limited band. This is achieved by including C6, which has a high reactance at low frequencies, and little affect on the circuit. At higher frequencies its reactance falls, and it virtually short circuits the track of VR1. This makes the setting of VR1 largely irrelevant, with the closed loop voltage gain being set at unity by R8 and R9. This restricts the filter to operation at bass frequencies.

The stage based on IC2b has lower capacitor values, which means that C7 only has a significant affect at frequencies of around 200Hz or more. C6 also comes into operation at a higher frequency of around 700Hz, and this stage therefore operates over a slightly higher frequency range. The capacitors in the other two stages are smaller by factors of about 4.5 and 20, which means that they operate over correspondingly higher frequency bands. VR1 to VR4 therefore control the bass, lower middle, upper middle, and treble ranges respectively.

The current consumption of the circuit is about 10 or 11 milliamps from a 12 volt supply. It will operate with supplies of between 9 and 15 volts, or up to 30 volts if C1 and C2 have high enough voltage ratings. The circuit has 'hum' filtering and should give good results if used with a reasonably well smoothed mains power supply unit.

Construction

The component and copper side views of the graphic equaliser board are provided in Figure 14.2 and 14.3. The board measures 78 holes by 18 copper strips. While this board is not particularly difficult to construct, its large width means that it is easy to 'lose your way', and due care should be taken to get everything in the right place. Also be careful not to omit any of the link-wires that are liberally scattered across the board.

Details of the hard wiring are shown in Figure 14.4, which must be used in conjunction with Figure 14.2.

It is not necessary to use screened leads, but avoid having leads longer than about 150 millimetres or so. I used phono sockets for SK1 and SK2, but you can of course use alternative types of connector if they are better suited to your particular set-up. VR1 to VR4 are rotary potentiometers on the prototype, but slider potentiometers are equally suitable. However, the slider variety seems to be relatively scarce these days, and they are also rather awkward to mount neatly on the front panel. Rotary potentiometers are the easier option.

This is a further example of a circuit that can be built into a larger piece of equipment (amplifier, stereo radio, etc.) if desired. For stereo operation two circuit boards are required, one for each stereo channel. The potentiometers can be dual gang types so that the two channels are adjusted in unison, but many users prefer to have separate controls for the two channels.

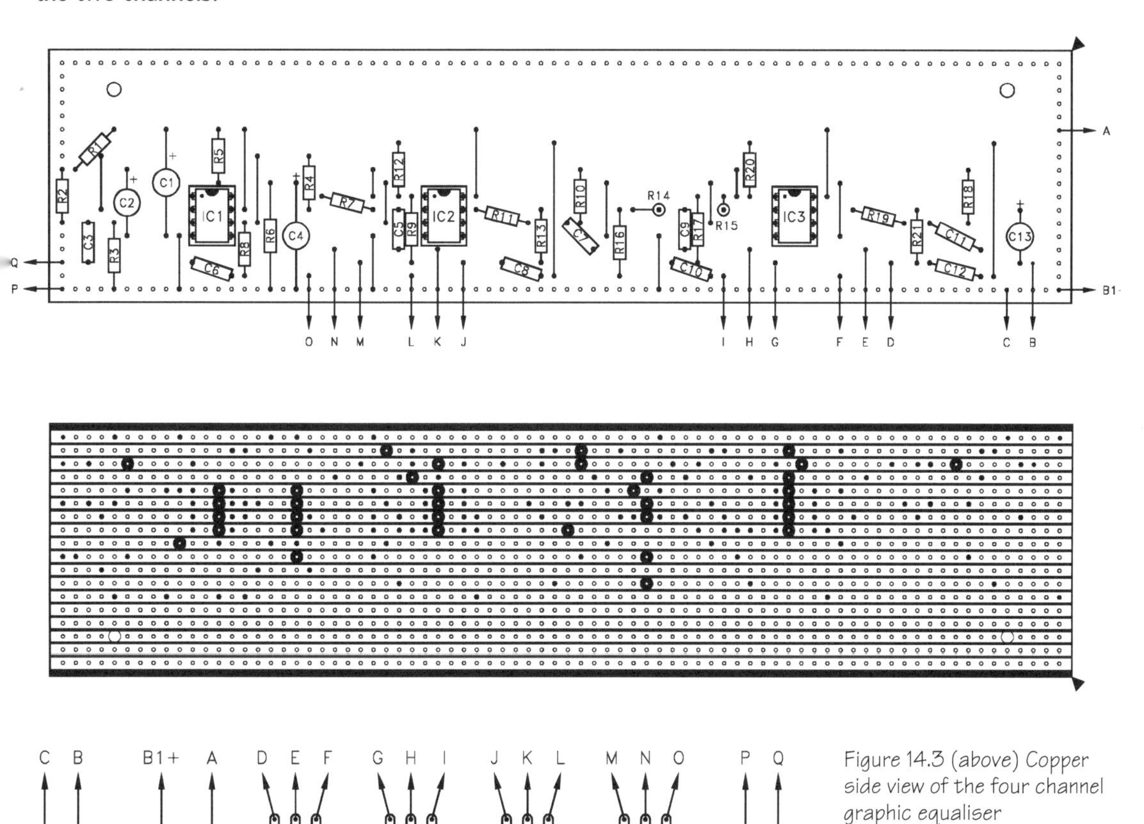

Figure 14.2 Component side view of the four channel graphic equaliser

Figure 14.3 (above) Copper side view of the four channel graphic equaliser

Figure 14.4 (left) Details of the hard wiring of the four channel graphic equaliser

In order to obtain a good signal to noise ratio the graphic equaliser must be used at a point in the system where it will receive a fairly high signal level. In other words, between (say) a tuner or an RIAA preamplifier and a power amplifier, and not between a microphone and a microphone preamplifier.

The effect of the unit should be very obvious with practically any programme material, but the filtering is most obvious on a noise source (such as the output from an f.m. tuner set between stations).

Components list

Resistors (all 0.25 watt 5% carbon film)

R1	2k2
R2,3	100k (2 off)
R4,7,10,11,14,15,18,19	2M2 (8 off)
R5,6	3k9 (2 off)
R8,9,12,13, 16,17,20,21	12k (8 off)

Potentiometers

VR1 to VR4	100k lin carbon (4 off)

Capacitors

C1,2	100u 16V radial elect (2 off)
C3	470n polyester, 7.5mm lead spacing
C4	47u 25V radial elect
C5	4n7 polyester, 7.5mm lead spacing
C6	47n polyester, 7.5mm lead spacing
C7	1n polyester, 7.5mm lead spacing
C8	10n polyester, 7.5mm lead spacing
C9	220p polystyrene
C10	2n2 polyester, 7.5mm lead spacing
C11	47p polystyrene
C12	470p polystyrene
C13	10u 25V radial elect

Semiconductors

IC1	TL071CP Bifet op amp
IC2,3	TL072CP dual Bifet op amp (2 off)

Miscellaneous

S1	S.P.S.T. min toggle switch
SK1,2	Phono socket (2 off)
B1	12 volt (e.g. 8 x HP7 size cells in holder)

Metal case, 0.1 inch pitch stripboard measuring 78 holes by 18 copper strips, control knob (4 off), battery connector, 8-pin d.i.l. holder

15

Audio mixer

An audio mixer is used to feed several signal sources into a single audio device, such as an amplifier or tape recorder. Virtually all mixers enable the level of each signal to be individually adjusted, as well as providing an overall level control.

A typical application for a mixer is when editing the audio track of a video tape. It is often desirable to add some background music and (or) a commentary to the original sound track. This is achieved by making a copy of the tape, but with the additional audio material being mixed with the original sound track during the copying process. There are many other applications for audio mixers though.

This mixer has four inputs, but it is easily expanded to practically any desired number of channels. The unit is described here as a mono mixer, but it can be built as a stereo unit if preferred. The input impedance is about 10k for all four inputs, and with the 'faders' at maximum there is a voltage gain of slightly less than two from each input to the output.

The circuit

Figure 15.1 shows the circuit diagram for the audio mixer. IC2 simply operates as a buffer stage at the output of the unit, and it is IC1 that provides the mixing action. It is used in a conventional summing mode mixer

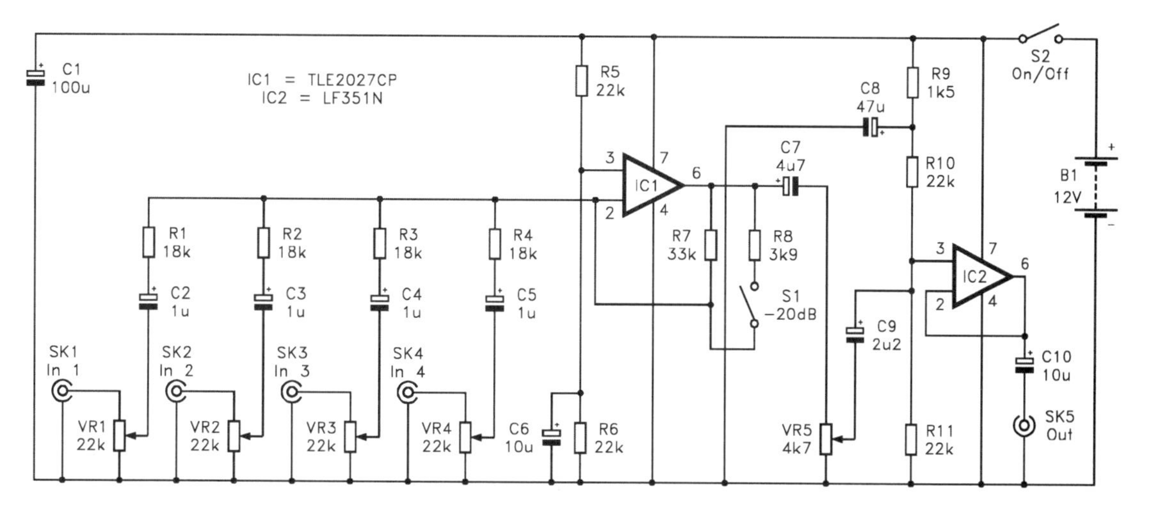

Figure 15.1 Circuit diagram for the audio mixer

circuit, which is a variation on the inverting amplifier mode. R5 and R6 bias the non-inverting input in the normal way, and R7 is part of the negative feedback network. In an inverting mode amplifier there is a second negative feedback resistor connected between the signal source and the inverting input of the operational amplifier. In a summing mode mixer circuit there are several of these resistors, one for each input signal. In this case the input resistors are R1 to R4. The circuit operates in essentially the standard inverting mode fashion, but with several input resistors and input signals the output has to balance the sum of the input voltages. This gives the required mixing action, with the mixed signal appearing at the output of IC1.

It is essential that a mixer circuit gives good isolation between the inputs. Otherwise there can be problems with the gain control for one channel affecting the gains of the other channels. Also, there would be a risk of one output driving significant amounts of current through one or more of the other outputs. The summing mode configuration provides almost total isolation between the inputs because what is termed a 'virtual earth' is formed at the input of IC1. This virtual earth is produced by the negative feedback action, which maintains the inverting input of IC1 at the mid-supply bias level. As a result of this, adjustment of one of the 'fader' controls (VR1 to VR4) has no significant affect on the other channels.

S1 can be used to provide an instant 'fade' by 20dB (i.e. the output level is reduced by a factor of about 10) by shunting R8 across R7. This can be useful in some applications, but S1 and R8 can be omitted if this is not a feature that is not applicable to your particular use for the mixer. VR5 is the overall level control. The current consumption of the circuit is approximately 5.5 milliamps.

Expansion

In order to add more channels it is merely necessary to use an extra 18k input resistor, 1u coupling capacitor, 22k 'fader' potentiometer, and input socket for each additional channel. This gives rise to the inevitable question, 'what is the maximum number of channels that can be used?' This is a 'how long is a piece of string?' type of question, with no clear-cut answer. Although IC1 is operating at a closed loop voltage gain of only about two, its effective voltage gain is much higher. The four input resistors are effectively connected in parallel, and having four inputs at a voltage gain of two is effectively the same as one input at a voltage gain of eight. Therefore, the more inputs that are added, the greater the strain on IC1. A larger number of inputs results in more noise, more distortion, and reduced closed loop bandwidth.

If large numbers of inputs are used (around 10 or 20), there is a definite advantage in using a high quality audio operational amplifier for IC1.

The TLE2027 specified for IC1 is a very high quality operational amplifier that provides very low levels of noise and distortion. It also offers a wide bandwidth with its gain bandwidth product of 13MHz. In fact it is perhaps a little over-specified for a four channel mixer, and except where very high performance is essential, a Bifet device such as an LF351N will suffice.

Two mixer circuits are required for stereo operation, one for use in each of the stereo channels. Battery B1 and on/off switch S2 are then common to both channels, and S1 would have to be a double pole switch with one pole used in each channel. The 'fader' potentiometers could be dual gang types so that the two channels are adjusted in unison, or separate controls, as preferred. A popular way of handling things is to have separate slider potentiometers, but with the potentiometers arranged in stereo pairs mounted side-by-side. This makes it easy to either adjust both channels in unison, or adjust one channel. One slight drawback of slider controls is that they are relatively awkward to mount in the case, and it can be difficult to produce really neat results. They are also relatively difficult to obtain these days (10k logarithmic slider potentiometers can be used if 22k components prove to be unobtainable).

Construction

A stripboard having 50 holes by 20 copper strips is required for the audio mixer project. The component layout, copper side view of the board, and hard wiring are shown in Figures 15.2, 15.3 and 15.4 respectively.

The board has been designed to make it easy to add extra inputs. It is just a matter of increasing the length of the board by five holes per additional input, and then adding the extra resistor and capacitor for each input following the general pattern of the existing input resistors and capacitors. It is not essential to used screened leads for the input wiring, but it is probably best to do so if the mixer has a large number of inputs. Otherwise there will be a vast amount of non-screened wiring which will leave the circuit vulnerable to stray pickup of electrical noise. The unit must be housed in a metal case that will give good screening, and help to avoid significant stray pick-up of 'hum' and other electrical noise.

In theory the unit should work reasonably well if it is used with low level input signal sources, such high impedance microphones. In practice this could give a comparatively low signal-to-noise ratio. The problem is partly that noise generated by the mixer itself will tend to degrade the signal-to-noise ratio, but the problem of stray pickup is also greatly

Photo 15.1 The finished audio mixer board

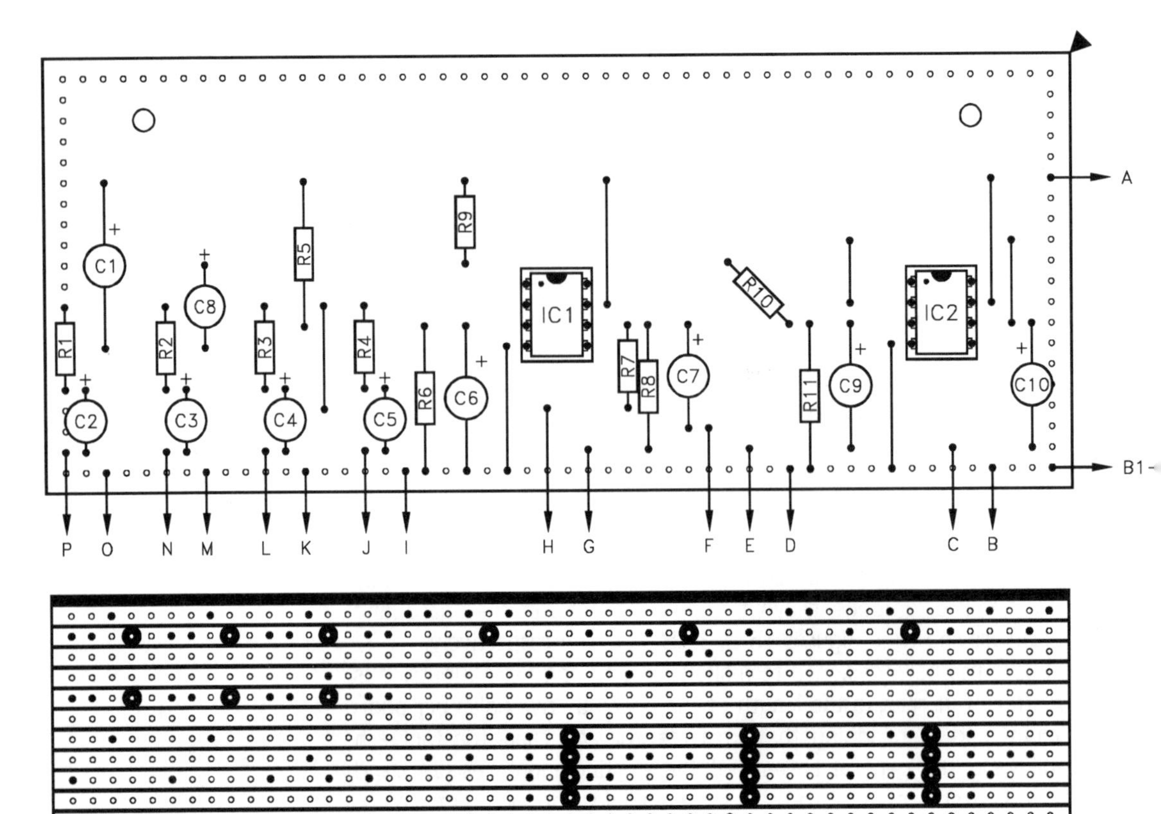

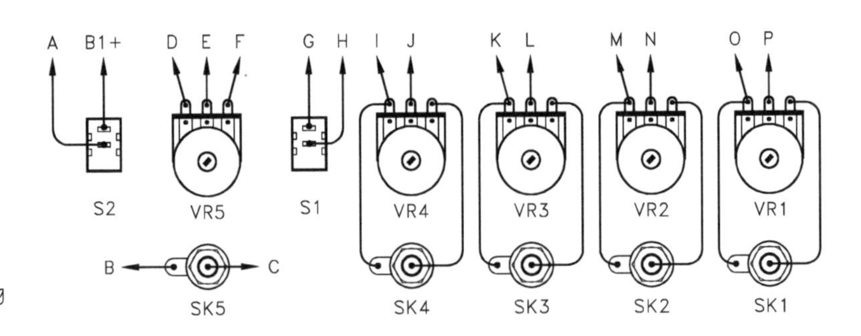

Top Figure 15.2 Component layout for the audio mixer board

Middle Figure 15.3 Copper side view of the board. The board has 50 holes by 20 copper strips

Right Figure 15.4 Hard wiring of the audio mixer

increased. It is obviously more complex and expensive to have a preamplifier ahead of each input of the mixer, rather than just one preamplifier at its output. However, having separate preamplifiers and the mixer operate at a much higher signal level will give much better results. Of course, by using various types of preamplifier ahead of some of the mixer's inputs it is possible to mix signals from a wide range of high and low level sources.

List of components

Resistors (all 0.25 watt 5% carbon film)

R1,2,3,4	18k (4 off)
R5,6,10,11	22k (4 off)
R7	33k
R8	3k9
R9	1k5

Potentiometers

VR1,2,3,4	22k log carbon rotary (4 off, see text)
VR5	4k7 log carbon rotary

Capacitors

C1	100u 16V radial elect
C2,3,4,5	1u 50V radial elect (4 off)
C6,10	10u 25V radial elect (2 off)
C7	4u7 50V radial elect
C8	47u 25V radial elect
C9	2u2 50V radial elect

Semiconductors

IC1	TLE2027CP very low noise op amp
IC2	LF351N Bifet op amp

Miscellaneous

B1	12 volt (e.g. 8 x HP7 size cells in holder)
S1,2	S.P.S.T. min toggle switch (2 off)
SK1 to SK5	Phono socket (5 off)

Metal case, 0.1in pitch stripboard measuring 50 holes by 20 copper strips, control knob (5 off), battery connector, 8-pin d.i.l. holder (2 off), wire, solder, etc.

16

Small audio power amplifier

A small audio power amplifier remains one of the most useful circuits, either for use in various electronic projects, or as a stand-alone unit for use as a general purpose amplifier for the electronics workshop. The circuit featured here operates over a supply voltage range of 4 to 12 volts (15 volts absolute maximum), and provides an output power of between about 100 and 600mW r.m.s. into an 8 ohm impedance loudspeaker. A quiescent current consumption of just 4 milliamps gives good battery life, but the unit will also work well when powered from a battery eliminator or other mains power supply unit.

The circuit

As can be seen from the circuit diagram of Figure 16.1, this circuit is based on a single integrated circuit. This is an LM386N, which contains a preamplifier and class B complementary output stage. At the input of the circuit C2 provides D.C. blocking and VR1 is a conventional volume control.

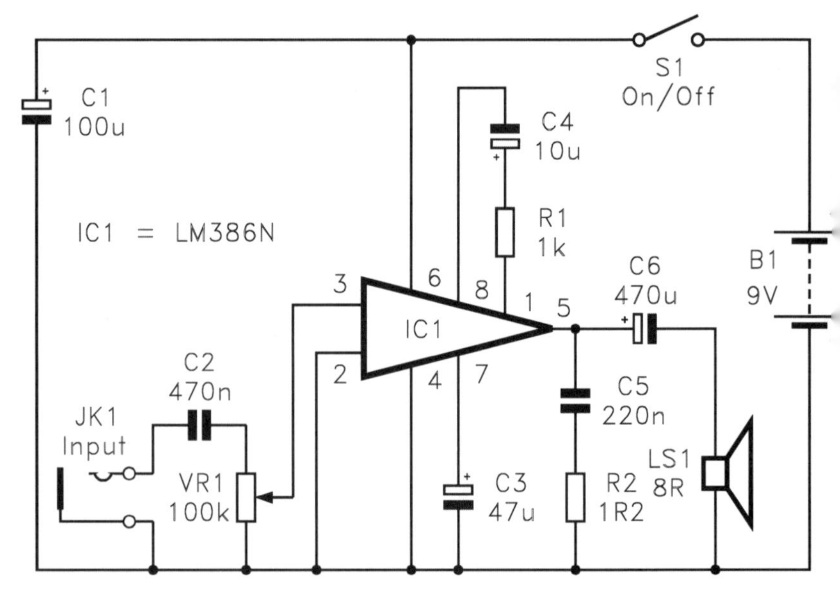

Figure 16.1 Circuit diagram of the small audio power amplifier

No coupling capacitor is needed between VR1 and the input of IC1, since the LM386N is a device which works quite happily with its input either 'floating' or referenced to the 0 volt supply rail. However, the lack of a capacitor at the input of IC1 makes it essential to have one ahead of VR1, or any D.C. component on the input signal could seriously unbalance IC1's biasing. IC1 actually has operational amplifier style inverting and non-inverting inputs at pins two and three respectively. In this case the differential capability of the device is of no value, and the input signal is applied to the non-inverting input. The inverting input is connected to ground to avoid stray pickup.

C1 is the main supply decoupling capacitor, and C3 decouples the supply to IC1's preamplifier stage. The voltage gain of the circuit is governed by the value of R1. With R1 and C4 omitted the voltage gain is only about 20, but this is adequate for some applications. With R1 at a value of zero the voltage gain is around 200, which is clearly quite high, but might be suitable for some applications. The specified value of 1k gives a voltage gain of approximately 45, which means that an input level of about 60 millivolts r.m.s. is needed in order to provide maximum output. C6 provides D.C. blocking at the output of the circuit. C5 and R2 are needed to aid good stability.

Although the standby current consumption of the circuit is only about four milliamps, at high volume levels the current drain can exceed 100 milliamps. It is therefore advisable to power the unit from a reasonably high capacity battery, such as six HP7 (AA) size cells. If a mains power supply unit is used, it must be capable of providing output currents of up to at least 100 milliamps, and should preferably have a current rating of 200 milliamps or more.

Construction and use

The small audio power amplifier is built on a stripboard which has 32 holes by 16 copper strips using the component layout shown in Figure 16.2. The copper side view of the board appears in Figure 16.3. This is a very simple board which offers nothing out of the ordinary, but make sure that all three link-wires are included. Details of the hard wiring is provided in Figure 16.4, which must be used in conjunction with Figure 16.2. This wiring is very straightforward, and it is not essential to use any screened leads.

While the audio quality provided by the LM386N falls some way short of hi-fi standards, it provides reasonable results. In most cases it is the reproduction quality of the loudspeaker that will be the limiting factor. The LM386N does not include either output short circuit or over-temperature protection. Short circuits and overloads will not necessarily prove to be 'fatal' for the LM386N, but it obviously makes sense to play safe and avoid them as far as possible. Do not use a loudspeaker having an impedance of less than eight ohms. Loudspeakers having higher impedances can be used, but will give reduced maximum output power.

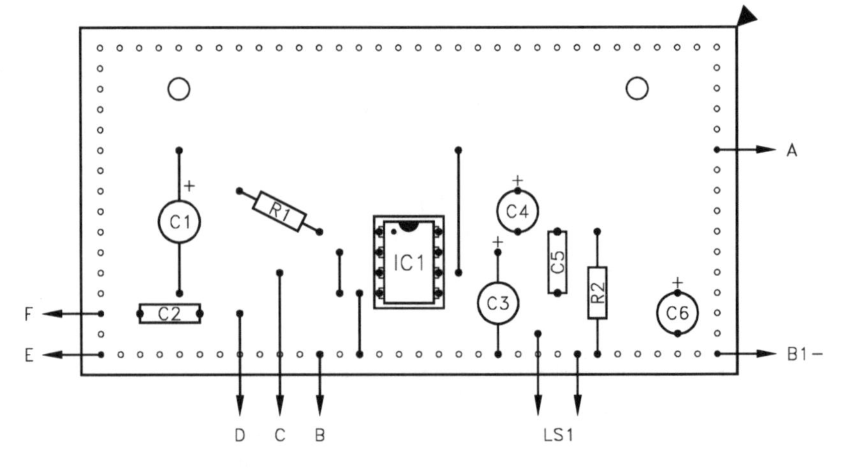

Figure 16.2 Component layout of the small audio power amplifier

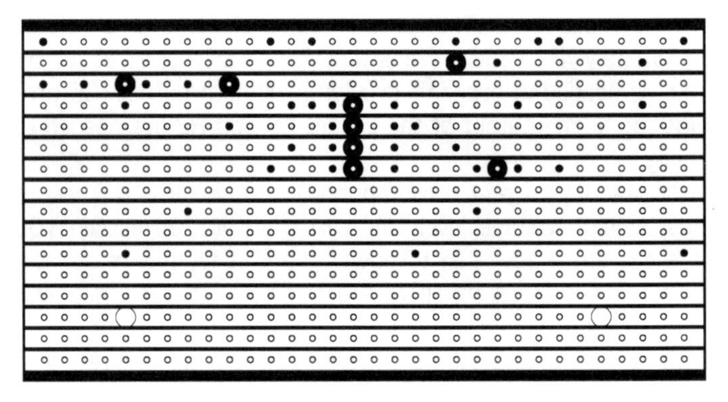

Figure 16.3 Copper side view of the board of the small audio power amplifier

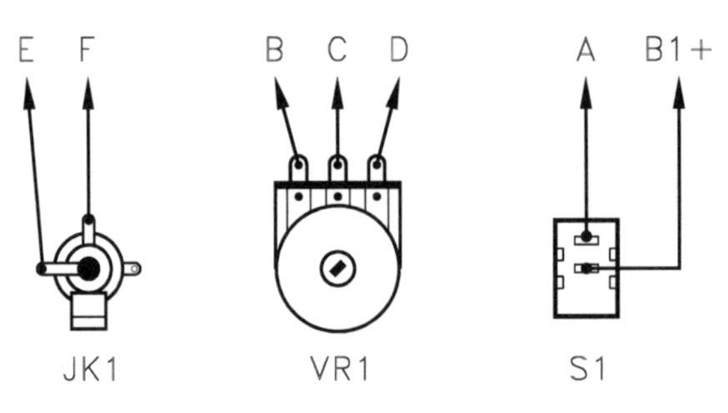

Figure 16.4 Small audio power amplifier hard wiring

List of components

Resistors (all 0.25 watt 5% carbon film)

R1	1k
R2	1R2

Potentiometer

VR1	100k log carbon rotary

Capacitors

C1	100u 16V radial elect
C2	470n polyester, 7.5mm lead spacing
C3	47u 25V radial elect
C4	10u 25V radial elect
C5	220n polyester, 7.5mm lead spacing
C6	470u 16V radial elect

Semiconductor

IC1	LM386N or LM386N-1 audio power amp

Miscellaneous

S1	S.P.S.T. min toggle switch
B1	9 volt (6 x HP7 size cells in holder)
JK1	3.5mm jack socket
LS1	Miniature 8 ohm impedance loudspeaker

Metal case, 0.1 inch pitch stripboard measuring 32 holes by 16 copper strips, 8-pin d.i.l. holder, battery connector, wire, solder, etc.

17

6 watt power amplifier

This simple but useful power amplifier module can provide output powers of up to 3 watts r.m.s. into an eight ohm impedance loudspeaker, or 6 watts r.m.s. into a four ohm type. It will operate over a supply voltage range of eight to 15 volts (18 volts absolute maximum), but a 15 volt supply is needed to provide the output powers mentioned above. While not in the true hi-fi category, the distortion level is about 0.1% or less at low and medium output powers. An input level of about 150 millivolts r.m.s. is needed to fully drive the amplifier, and the input impedance is 4k7.

The circuit

The circuit diagram for the 6 watt power amplifier appears in Figure 17.1. It is based on a TDA2003 audio power amplifier chip which is a form of operational amplifier. Some integrated circuit power amplifiers are actually high power versions of operational amplifiers, and are used in standard operational amplifier configurations. The TDA2003 is essentially a high power version of an operational amplifier, but it has built-in biasing components, and it has to be used in circuit configurations that take this into account.

VR1 is the volume control, and the output from its wiper is coupled to the non-inverting input of IC1 by way of C3. The closed loop voltage gain of IC1 is controlled by R1 and R2 which form a conventional negative

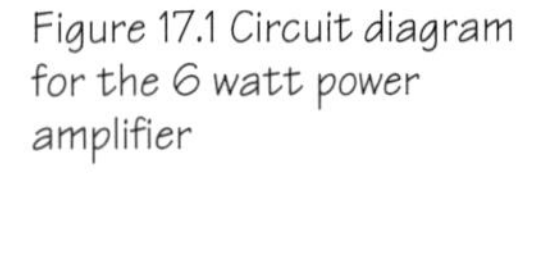
Figure 17.1 Circuit diagram for the 6 watt power amplifier

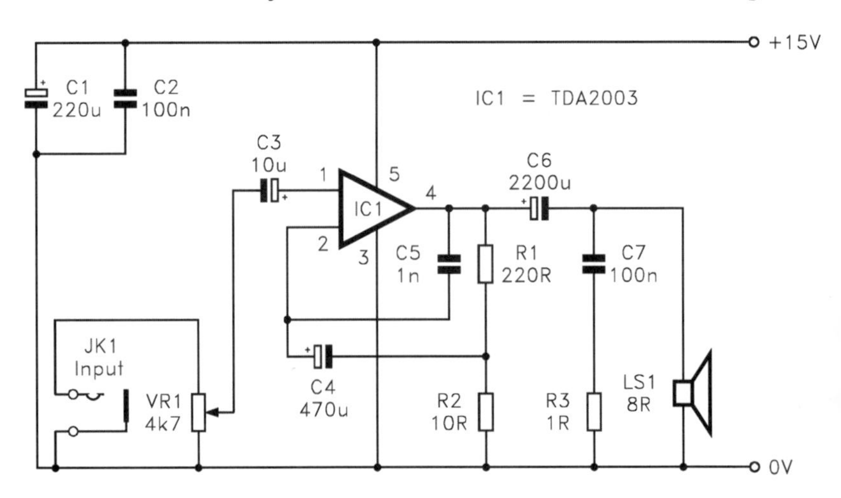

feedback circuit. They set the voltage gain in the normal way, giving a gain figure of 23 with the specified values. If necessary, the voltage gain can be boosted slightly by making R2 a little lower in value (but no less than 2R2). C4 provides D.C. blocking between the negative feedback circuit and the inverting input of IC1. C5 provides increased feedback at high frequencies which aids good stability. C7 and R3 are also needed to ensure that there are no problems with instability. C6 provides D.C. blocking at the output.

The quiescent current consumption of the circuit is around 45 milliamps, but the supply current can increase to around 300 milliamps when using an eight ohm loudspeaker, or 600 milliamps when using a four ohm type. Battery operation is only practical if high capacity cells are used (e.g. 10 HP2/D size cells), and the use of rechargeable batteries is preferable. Of course, a mains power supply unit can be used provided it has suitable voltage and current ratings, and a reasonably well smoothed output. Good results are obtained using a 12 volt regulated battery eliminator rated at 500 milliamps or more, but it is preferable to use a four ohm impedance loudspeaker in order to obtain a reasonably high output power (about four watts r.m.s. as opposed to just over two watts r.m.s. with an 8 ohm loudspeaker).

The amplifier will also work well with a 12 volt battery eliminator that does not have a regulated output, and is rated at about 700 to 800 milliamps. The loaded output voltage will be significantly more than 12 volts, which is reflected in higher output powers than those obtained using a stabilised battery eliminator. The quiescent supply voltage should be just within the maximum 18 volt supply potential of the TDA2003. A non-regulated battery eliminator would seem to be the best choice in this case, as it offers more output power for what is normally a slightly lower cost.

Construction

Details of the stripboard layout and wiring are provided in Figure 17.2. The illustration for the underside of the board is provided in Figure 17.3. A stripboard measuring 37 holes by 23 copper strips is required. The only unusual aspect of construction is that IC1 must be provided with a heatsink.

The TDA2003 is actually available in two versions, which differ only in that the leadout wires are pre-formed for either horizontal or vertical mounting. The type number in component catalogues often has a 'H' or 'V' suffix to distinguish between the two types, but this suffix letter is not present on any of the actual devices that I have seen. In this case either type is acceptable, but I would recommend using the horizontal mounting version. This can be bolted to the circuit board, which gives a decidedly tougher finished product. The heatsink fits between IC1 and the board, and is held in place by the same bolt that secures IC1 to the circuit board. A heatsink having a rating of about 10 degrees Celsius per watt should be adequate. In other respects the board and wiring are perfectly straightforward. The TDA2003 has both output short circuit and thermal overload protection incidentally.

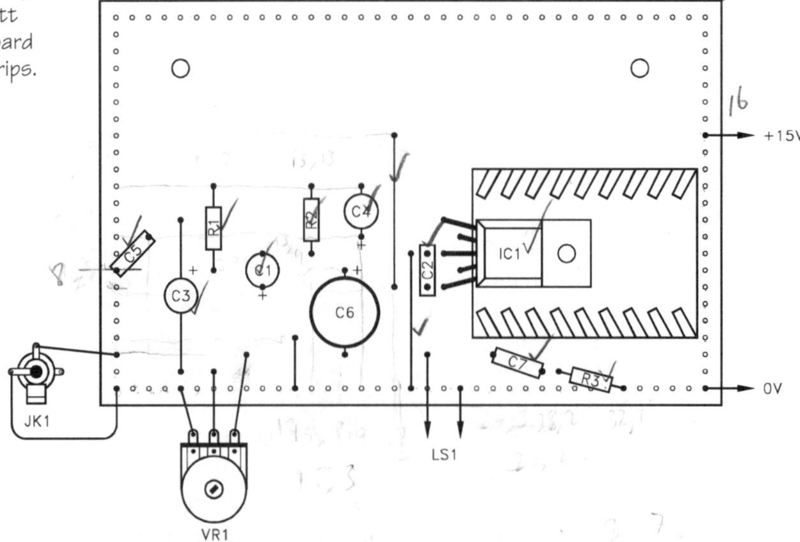

Figure 17.2 Stripboard layout and wiring of the 6 watt power amplifier. The board has 37 holes by 23 strips.

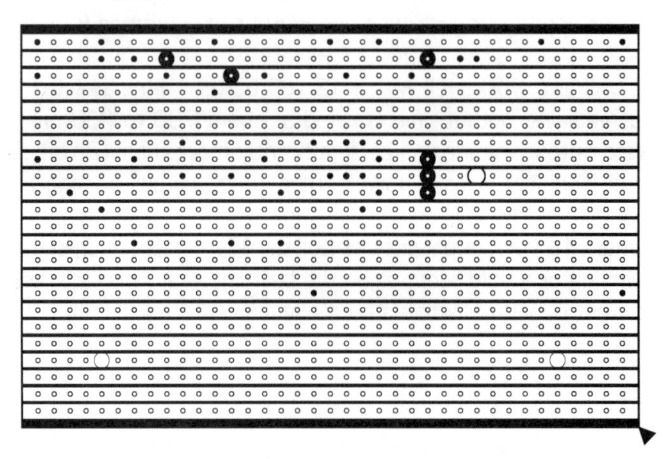

Figure 17.3 Underside of the board of the 6 watt power amplifier

List of components

Resistors (all 0.25 watt 5% carbon film)
R1 220R
R2 10R
R3 1R

Potentiometer
VR1 4k7 log carbon rotary

Capacitors
C1 220u 25V radial elect
C2 100n ceramic

C3 10u 25V radial elect
C4 470u 25V radial elect
C5 1n polyester, 7.5mm lead spacing
C6 2200u 16V radial elect
C7 100n polyester, 7.5mm lead spacing

Miscellaneous
LS1 4/8 ohm impedance loudspeaker, capable of handling at least 3/6 watts r.m.s.
JK1 Standard jack socket
Metal case, 0.1 inch stripboard measuring 37 holes by 23 copper strips, finned bolt-on heatsink having a rating of about 10 degrees Celsius per watt, control knob, wire, solder, etc.

18

20/32 watt power amplifier

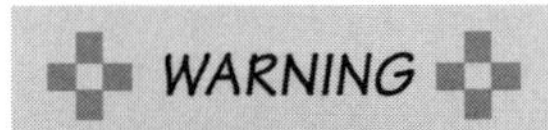

WARNING

The mains supply is extremely dangerous, and projects which connect to the mains are only suitable for those who have the necessary project building experience and expertise

This power amplifier module will provide output powers of up to about 20 watts r.m.s. into an eight ohm impedance loudspeaker. Basing the unit on a modern power amplifier integrated circuit results in an extremely simple circuit, but the performance of the unit is quite good. The typical harmonic distortion is under 0.1% at most power levels, making the circuit suitable for all but the most demanding of applications.

Some early audio power amplifier integrated circuits suffered from a high noise level, and were notoriously unstable. The audio power amplifier chip used in this design has a negligible noise level and achieves good stability without having to resort to elaborate printed circuit designs liberally scattered with 'compensation' components.

The circuit

The circuit diagram for the 20 watt audio power amplifier appears in Figure 18.1. The unit is based on a TDA2050 power amplifier, which takes the same physical form as the TDA2003 used in the 6 watt amplifier featured previously. In other words, it is a plastic power device which has five leadout wires. However, it is somewhat different to the TDA2003 in that it does not have any built-in bias or feedback compo-

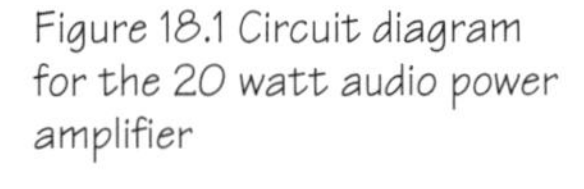

Figure 18.1 Circuit diagram for the 20 watt audio power amplifier

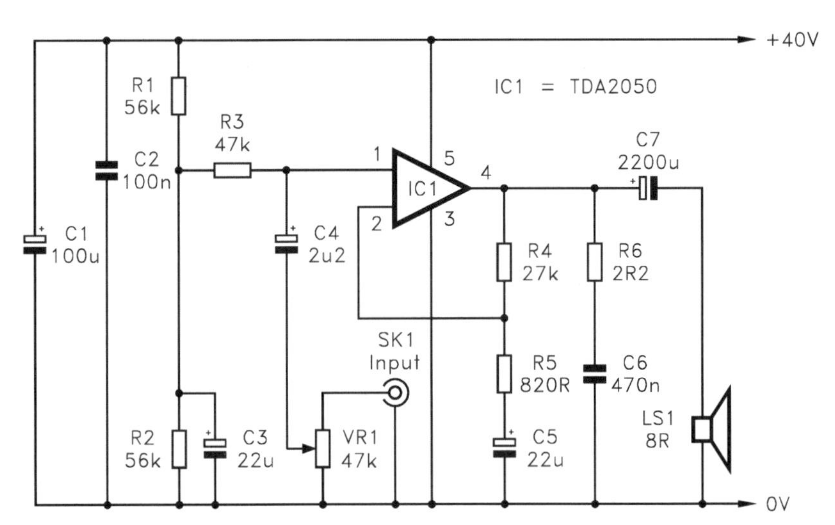

nents. It is effectively an ordinary operational amplifier, but with a high power class B output stage that can deliver output powers of more than 30 watts r.m.s. It has internal thermal overload and short circuit protection circuits incidentally.

Being an operational amplifier, the TDA2050 can be used in the standard inverting and non-inverting amplifier modes. In this case it is used as a non-inverting amplifier. The non-inverting input is biased by R1, R2, and R3. The potential divider formed by R1 and R2 produces an output potential of half the supply voltage, and this is coupled to the non-inverting input of IC1 by way of R3. C3 decouples any 'hum' or other noise on the supply lines, so that it is not fed to the non-inverting input of IC1 via the bias circuit. R3 sets the basic input impedance of the amplifier at 47k, but the shunting effect of volume control VR1 reduces the overall input impedance to 23.5k at the maximum volume setting.

R4 and R5 are the negative feedback network, and they set the closed loop voltage gain of the circuit at about 34. Some 13 volts r.m.s. is needed at the output of the circuit in order to produce maximum output power, which means that an input level of about 380 millivolts r.m.s. is needed to fully drive the circuit. C5 provides D.C. blocking in the feedback circuit, and results in unity voltage gain at D.C. The output of the amplifier is therefore biased to the same half supply potential as the non-inverting input. This enables the highest possible output level to be achieved before the onset of clipping and serious distortion. R6 and C6 are needed to aid good stability, and C7 provides D.C. blocking at the output.

The circuit requires a supply potential of about 40 volts or so in order to achieve an output power of 20 watts r.m.s. into an eight ohm load. It will actually provide some 25 watts r.m.s. from a 45 volt supply, but it has to be borne in mind that the absolute maximum voltage rating for the TDA2050 is 50 volts. Realistically, it would require a stabilised supply in order to obtain 25 watts r.m.s., whereas 20 watts can be provided by a non-stabilised supply.

The salient point here is that the class B output stage of the TDA2050 results in large variations in supply current. The typical supply current is only 55 milliamps under quiescent conditions, but the consumption can exceed ten times this figure at high output powers. This results in a significant reduction in the output voltage of a non-stabilised supply at high output powers. This places a practical limit on the maximum loaded supply potential that can be achieved without exceeding the 50 volt maximum under quiescent conditions.

The TDA2050 can be used with a four ohm impedance loudspeaker incidentally, and output powers of up to about 32 watts r.m.s. can then be achieved. However, this obviously boosts the current consumption of the circuit quite substantially. It also increases the dissipation in IC1, so that a much larger heatsink is required. The prototype amplifier has only been fully tested with eight ohm impedance loudspeakers.

Construction

The stripboard layout and wiring for the 20 watt amplifier are provided in Figure 18.2. The underside view of the board is shown in Figure 18.3, but no breaks are needed in the copper strips. The board measures 26 holes by 26 copper strips.

The TDA2050 is available with the leadout wires pre-formed for vertical or horizontal mounting (usually indicated respectively by 'V' and 'H' suffixes to the type numbers in component catalogues). The vertical version is used in the prototype amplifier, but there should be no difficulty in

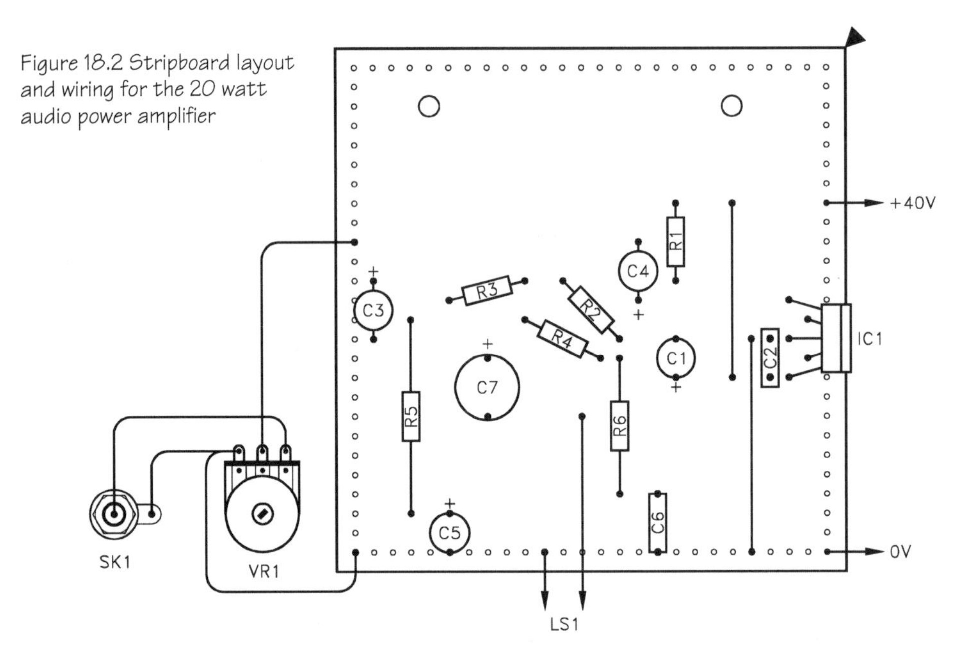

Figure 18.2 Stripboard layout and wiring for the 20 watt audio power amplifier

Figure 18.3 Underside view of the board. There are no breaks in the strips.

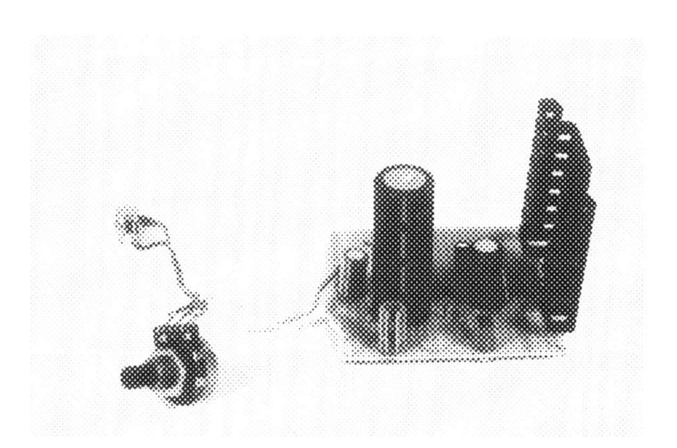

Photo 18.1 The completed 20/32 watt power amplifier

using either type with this board layout. The leadout wires of IC1 need to be splayed very slightly in order to fit the device onto the circuit board.

IC1 has to dissipate significant amounts of power, especially when the amplifier is used at high volume levels. A large bolt-on heatsink having a rating of about six to seven degrees Celsius per watt should just about suffice provided the amplifier is housed in a case which has generous ventilation slots, and is only used with eight ohm loudspeakers. However, it is better to use a larger chassis mounted heatsink, or in some instances it might be possible to use the metal housing itself as the heatsink. The metal heat-tab of IC1 connects internally to the 0 volt supply rail, but this should not give any problems since the chassis of the amplifier will normally be connected to the 0 volt supply as well. Therefore, it should not be necessary to use an insulating kit when mounting IC1 on the heatsink.

List of components

Resistors (all 0.25 watt 5% carbon film)

R1,2	56k (2 off)
R3	47k
R4	27k
R5	820R
R6	2R2

Potentiometer

VR1	47k log carbon rotary

Capacitors

C1	100u 63V radial elect
C2	100n ceramic
C3,5	22u 63V radial elect (2 off)
C4	2u2 63V radial elect
C6	470n polyester, 7.5mm lead spacing
C7	2200u 35V radial elect

Semiconductor
IC1 TDA2050 (see text)

Miscellaneous
LS1 8 ohm impedance loudspeaker capable of handling at least 20 watts r.m.s.
SK1 Phono socket
Case, 0.1 inch pitch stripboard having 26 holes by 26 copper strips, control knob, wire, solder, etc.

19

20/32 watt amplifier power supply

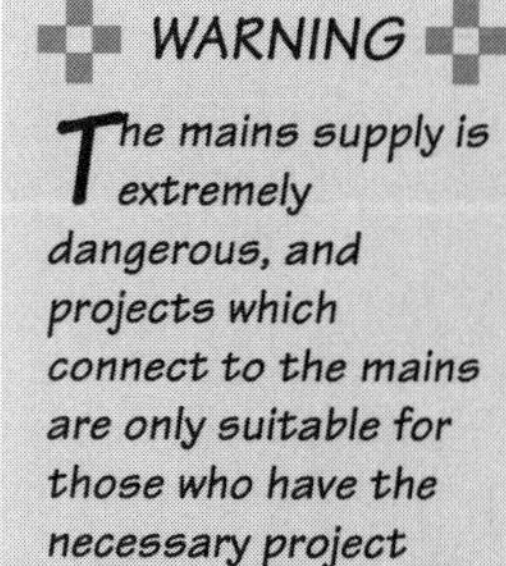

Battery operation is not really a practical proposition with the 20/32 watt amplifier, as it requires a relatively high supply voltage and consumes quite high supply currents. A mains power supply unit is required, and it is not necessary to use an elaborate circuit having a highly stable, low noise output. In fact a simple non-stabilised supply will suffice, although the output must be reasonably well smoothed if a low 'hum' level is to be obtained from the amplifier. A suitable power supply circuit is provided in Figure 19.1. S1 is the on/off switch, and from here the mains supply connects to the primary winding of step-down and isolation transformer T1. The output from the secondary winding of T1 is full-wave rectified by D1 to D4, which form a bridge rectifier. Smoothing is provided by C1, and fuse FS1 protects the circuit if an overload should occur.

Power supply circuits seem to produce a certain amount of confusion because the rating of the mains transformer never matches up with the output voltage and maximum output current rating of the supply. In this case a 30 volt mains transformer with a current rating of two amps is being used to produce a 45 volt supply with a maximum current drain that will probably not be much more than one amp. The first point to bear in mind here is that the 30 volt rating of the mains transformer is its r.m.s. a.c. rating, and not the D.C. potential that it will produce after rectification and smoothing. Multiplying 30 by 1.41 gives the peak output voltage of T1, but a volt or two has to be deducted from this figure to allow for the voltage drop through the rectifiers. In theory this gives

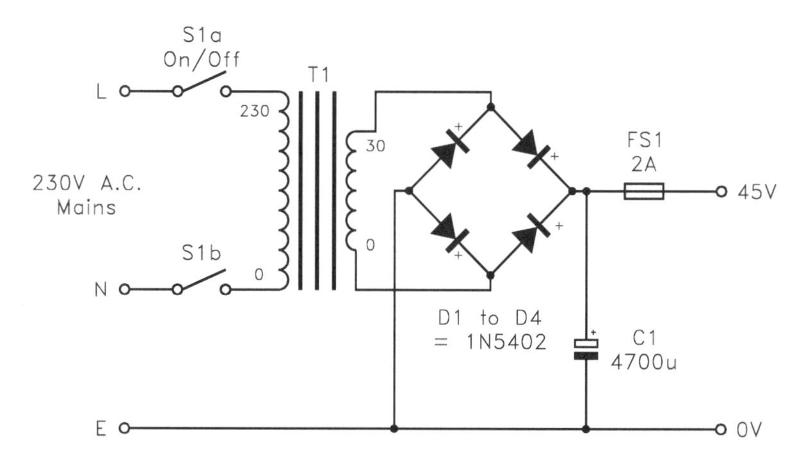

Figure 19.1 The circuit diagram for the mains power supply unit.

about 40 volts or so across C1, but under the low quiescent loading of the power amplifier the actual voltage will be significantly higher than this. It will typically be about 45 volts or a little more. However, at high output powers the current consumption of the amplifier increases substantially, and the output voltage of the supply reduces to about 40 volts.

The current rating of a mains transformer is its A.C. current rating, and the D.C. output current that can be taken after rectification and smoothing is much less. A bridge rectifier is the most efficient type, but it still only permits a d.c. output current that is about 62% of the mains transformer's A.C. current rating. Hence T1 must have a current rating of two amps in order to produce a D.C. output current of little more than one amp. There is actually some advantage in using a transformer having a higher current rating. The higher the current rating of T1, the less the output voltage will 'sag' under loading, and the higher the output power available from the amplifier. If you wish to obtain the highest possible output power from the amplifier, T1 should have a current rating of three amps or more. If the amplifier will be used with a four ohm impedance loudspeaker, three amps is the absolute minimum acceptable current rating for this component. FS1 should then have a rating of three amps.

It now seems to be the norm for mains transformers to have twin secondary windings, or a centre-tapped secondary winding. Therefore, T1 might have to be a twin 15 volt transformer with the secondary windings wired in series, or a 15 – 0 – 15 volt type with the 0 volt centre-tap ignored.

Construction

Although this project is extremely simple, neither it nor the 20/32 watt amplifier are really suitable for beginners. The mains supply is extremely dangerous, and projects which connect to the mains are only suitable for those who have the necessary project building experience and expertise. The standard safety measures must be implemented when building this project. While it is fine to experiment with temporary uncased 'lash-ups' when projects are powered from batteries or battery eliminators, this is not acceptable with a mains power supply unit. The unit must be housed in a strong metal case which should have a screw fitting lid or cover, rather than one which simply clips on and off. This ensures that there is no easy access to the dangerous mains wiring. The case must be reliably earthed to the mains earth lead.

There is no need to use a circuit board for this project, and it can be put together using hard wiring. The wiring diagram for the power supply is shown in Figure 19.2. I used a rotary on/off switch, and it is a switch of this type that is shown in Figure 19.2. Any d.p.s.t. switch can be used, provided it is rated for use on the 230 volt A..C. mains supply. Other types of switch will almost certainly have a different tag arrangement though, so check them with a continuity tester if you are unsure about the right method of connection. FS1 is mounted in a chassis mounting fuse-holder in the prototype power supply, but a panel mounting fuse-

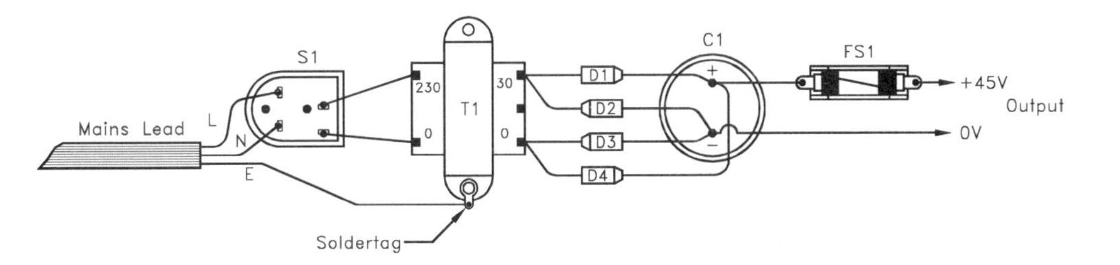

Figure 19.2 Wiring diagram for the mains power supply unit

holder is also suitable. C1 is a large 'can' type electrolytic which is mounted on the base panel of the case using a matching mounting clip.

A solder-tag fitted on one of T1's mounting bolts provides a chassis connection point for the mains earth lead. Make quite sure that this connection is physically strong (i.e. give the earth lead a good tug to see if it pulls away from the soldertag). D1 to D4 could be mounted on a circuit board or tag board, and then wired to T1 and C1, but it is probably easier and better to simply wire them directly between T1 and C1. However, the leads of four rectifiers should be fitted with p.v.c. sleeving to ensure that they can not accidentally short circuit to one another. Make quite sure that C1 and all four rectifiers are connected with the right polarity, as mistakes here could cause costly damage, and could even be dangerous. The rectifiers should fit into place quite easily provided the ends of their leadout wires and the tags of T1 and C1 are tinned with solder prior to making the soldered connections.

List of components

D1 to D4	1N5402 200v 3A rectifiers (4 off)
T1	Standard mains primary, 30 volt 2 amp secondary (see text)
C1	4700u 50V 'Can' electrolytic with mounting clip
S1	Rotary D.P.S.T. mains switch
FS1	20mm 1.5A 'quick-blow' fuse (see text)

Metal case, control knob, mains lead fitted with 2A fuse, 20mm chassis mounting fuse-holder, wire, solder, etc.

20

Dynamic noise limiter

Digital recording techniques and modern low noise semiconductors have greatly reduced the problems with background 'hiss' that dogged the audio world in the past. Unfortunately, digital sound technology has yet to permeate all aspects of sound recording, and the audio quality of many video recorders and camcorders is far from 'hiss' free. Also there are a great many cherished older recordings in most collections that suffer from excessive background 'hiss.' This is certainly a problem with many older cassette recordings.

Simple lowpass filtering can reduce background 'hiss' quite effectively, but it also attenuates the high frequency components in the main signal. The noise is reduced, but cymbal 'crashes' lack impact, as do many crescendos in the reproduced music. Without resorting to highly advanced digital sound processing techniques it is not possible to 'have your cake and eat it', with the noise being reduced and the main signal being left largely intact. However, dynamic analogue signal processing does offer an in-between approach which gives good 'hiss' reduction while still providing plenty of impact on crescendos.

Dynamic noise reduction relies on the fact that loud sounds tend to mask quieter sounds. The effectiveness of this masking depends on the relative frequency contents of the two sounds. The masking is most effective when the two sounds have similar frequency components. In order words, a high pitched sound is good at masking quieter high pitched sounds, but is much less effective at masking low frequency sounds.

A dynamic noise limiter is a form of lowpass filter, but it does not have a fixed cut-off frequency. The cut-off frequency can vary between about 5kHz and 20kHz, and it is controlled automatically. The stronger the input signal, the higher the cut-off frequency. This system provides a high level of noise reduction at low volume levels where the noise is most noticeable, and little or no noise reduction at higher signal levels where the noise is not perceptible anyway. At low volumes the reproduced signal will be somewhat lacking in treble content, but at high volume levels the full treble content is present, giving convincing cymbal crashes, orchestral crescendos, etc. Subjectively, the reduction in noise is quite high at around 12dB or more, but the loss of signal quality is quite small, giving an overall improvement in the reproduction quality. Although one might expect the changes in the operating frequency of the filter to be clearly audible, in practice things do not happen this way. Provided the cut-off frequency of the filter is varied, rather than having a fixed cut-off frequency and a varying attenuation rate, the changes in the filtering are not apparent to the listener.

System operation

The block diagram of Figure 20.1 shows the general arrangement used in this dynamic noise limiter. The main signal path is through a two stage voltage controlled filter (v.c.f.) which gives an overall attenuation rate of 12dB per octave. A buffer amplifier between the two filters ensures that there is no interaction between them, and a second buffer stage provides the unit with a suitably low output impedance.

The rest of the circuit consists of a side-chain that provides the control voltage for the v.c.f. The first stage in the side-chain is an amplifier which slightly boosts the signal. This feeds into a 3rd order (18dB per octave) highpass filter which has its cut-off frequency at about 1kHz. As pointed out previously, high frequency sounds mask other high frequency sounds very effectively, but low frequency sounds are ineffective at masking high frequency sounds. When applied to noise reduction this means that sounds which have a strong high frequency content will obscure the background 'hiss', but low frequency sounds will fail to do so, no matter how strong they are. If low frequency sounds were allowed to control the filter, the noise would be clearly heard to increase during strong low frequency signals (a phenomenon known as 'breathing effects'). The highpass filter ensures that the cut-off frequency of the v.c.f. will not be lifted by strong bass signals.

The output signal from the highpass filter is rectified and smoothed to produce a positive D.C. output voltage that is roughly proportional to the peak input level. This signal is boosted slightly by a D.C. amplifier, and it is then mixed with a variable D.C. potential before being fed to the control input of the v.c.f. The control which provides the variable D.C. bias

Figure 20.1 Block diagram of the dynamic noise limiter

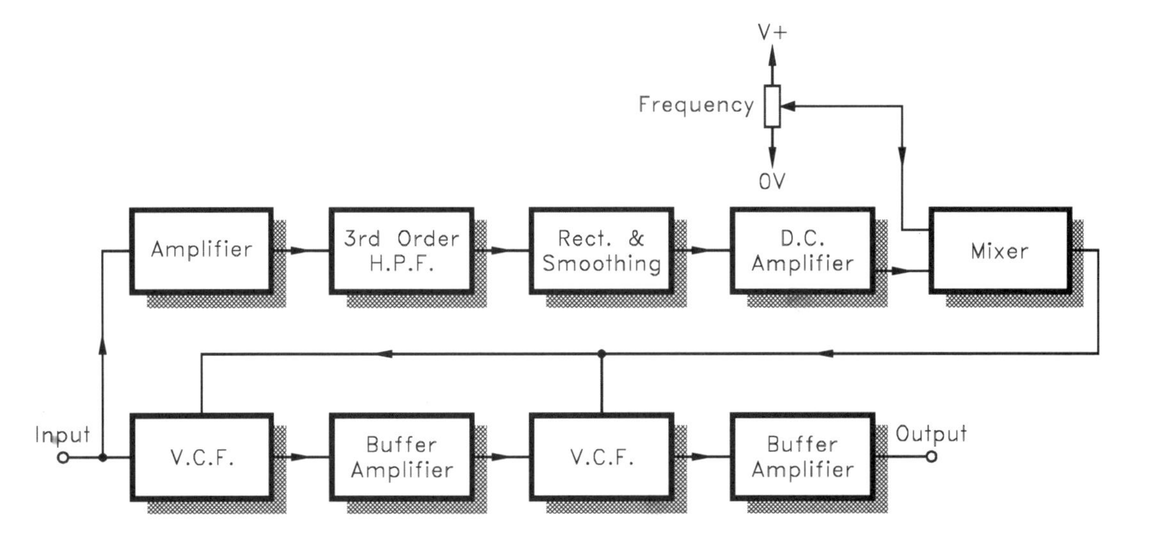

sets the minimum cut-off frequency of the v.c.f. In practice this is set to give the required amount of noise reduction, but the audio quality during periods of low volume will obviously suffer if it is set for too low a cut-off frequency. With low input levels the output voltage from the D.C. amplifier is very small, and it does not significantly raise the operating frequency of the v.c.f. If the input level is increased, the output voltage from the D.C. amplifier rises, and so does the filter's cut-off frequency. This gives the required automatic control of the v.c.f.'s cut-off frequency, and effective dynamic noise limiting.

The circuit

Refer to Figure 20.2 for the full circuit diagram for the dynamic noise limiter. The voltage controlled filter is based on the LM13700N dual transconductance operational amplifier, or the virtually identical LM13600N. A transconductance operational amplifier is very different to an ordinary operational amplifier, and the main difference is that a transconductance amplifier is current rather than voltage operated. Of more importance in most applications, a transconductance amplifier has an additional input, and the output current is a determined by the differential input current, and the bias current fed to the control input. This effectively gives an amplifier having an output impedance that can be controlled by an input current. Adding a resistor in series with the bias input converts the circuit from current to voltage control, and adding a capacitor across the output produces a simple voltage controlled lowpass filter.

The two sections of the LM13700N (IC1) are used in the same basic v.c.f. configuration. R1 and R2 provide a mid-supply bias voltage for the filters. Most of the resistors in this part of the circuit are concerned with biasing, or act as negative feedback networks which limit the maximum voltage gain through the filters to unity. Two exceptions are R7 and R11, which act as load resistors for the two internal buffer amplifiers of IC1. C4 and C5 are the two filter capacitors, And R10 is the series resistor which converts the filter from current to voltage control. VR1 is the potentiometer which controls the minimum cut-off frequency of the filter.

Some of the input signal is fed to IC2, which acts as a non-inverting amplifier having a closed loop voltage gain of about 5.5. The output of

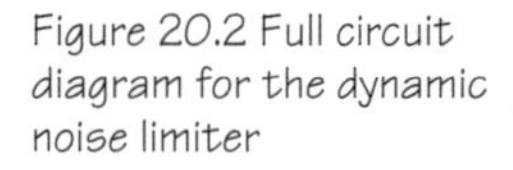
Figure 20.2 Full circuit diagram for the dynamic noise limiter

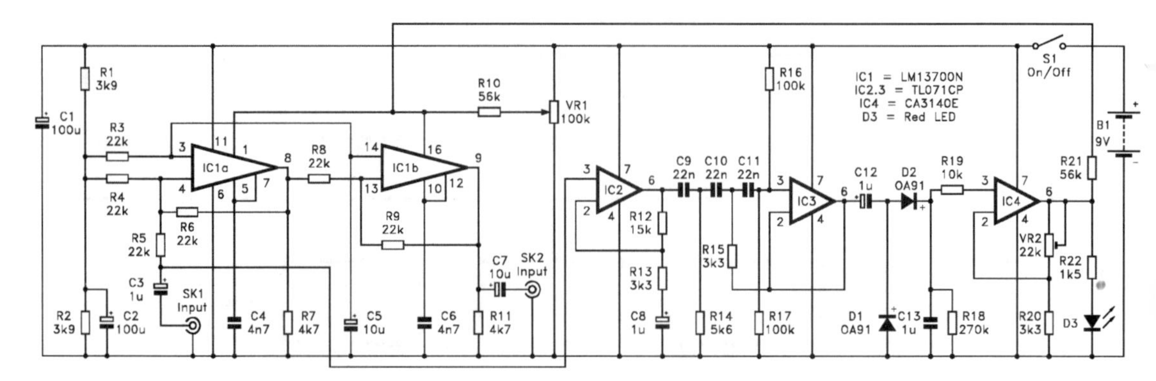

IC2 drives a conventional three stage active highpass filter based on IC3. This has its cut-off frequency at about 1kHz, which is high enough to ensure that strong low frequency signals do not give problems with breathing effects. On the other hand, it is low enough to ensure that there is sufficient high frequency output to operate the v.c.f. D1 and D2 rectify the output signal from IC3, and C13 provides smoothing. The circuit has a fast attack time so that it responds almost at once to increases in the signal level. The decay time has to be something of a compromise. Ideally it would be very short so that the circuit responds very rapidly to falls in the signal level. However, making the decay time too short results in distortion, and it is essential to use a relatively long decay time. With the specified values the distortion level is kept low, but sudden drops in the signal level are not accompanied by bursts of 'hiss.'

The D.C. amplifier uses IC4 in the standard non-inverting configuration. Note that IC4 is used without a negative supply rail, and that few operational amplifiers can function properly in single supply rail D.C. amplifier circuits. Consequently, devices such as the LF351N, TL071CP, uA741C, etc. will not work properly in the IC4 position of this circuit. VR2 enables the closed loop voltage gain of IC4 to be varied from unity to about 7.7, which enables the circuit to work properly with a reasonable range of input levels. The output from IC4 is coupled to the control input of the v.c.f. by way of R21. R10 and R21 form a simple passive mixer which combines the output of the side-chain circuitry and the frequency control potentiometer.

D3 is a l.e.d. indicator which is driven from the output of IC4 via current limiting resistor R22. D3 lights up when there is a strong high frequency content on the input signal and the v.c.f.'s cut-off frequency is being lifted. The stronger the high frequency components in the input signal, the brighter D3 lights up. This is useful when initially testing the unit, but is of little value thereafter. It is up to the constructor to choose whether or not to include this as a permanent part of the unit.

The circuit requires a supply voltage of about nine to 12 volts. The current consumption is about eight milliamps, but rises by a few milliamps when D3 lights up. A PP3 size nine volt battery is just about adequate to power the circuit, but a higher capacity battery will be more economic if the unit will receive a great deal of use. Alternatively, any regulated nine or 12 volt battery eliminator should give good results with this project. Two circuits are needed for stereo processing, one for use in each channel. This does, of course, double the current consumption of the unit, making a high capacity battery or a battery eliminator the only practical options.

Construction

A stripboard having 74 holes by 21 copper strips is used for this project. The component layout is shown in Figure 20.3, and the underside view of the board appears in Figure 20.4. The hard wiring is shown in Figure 20.5, which should be used in conjunction with Figure 20.3.

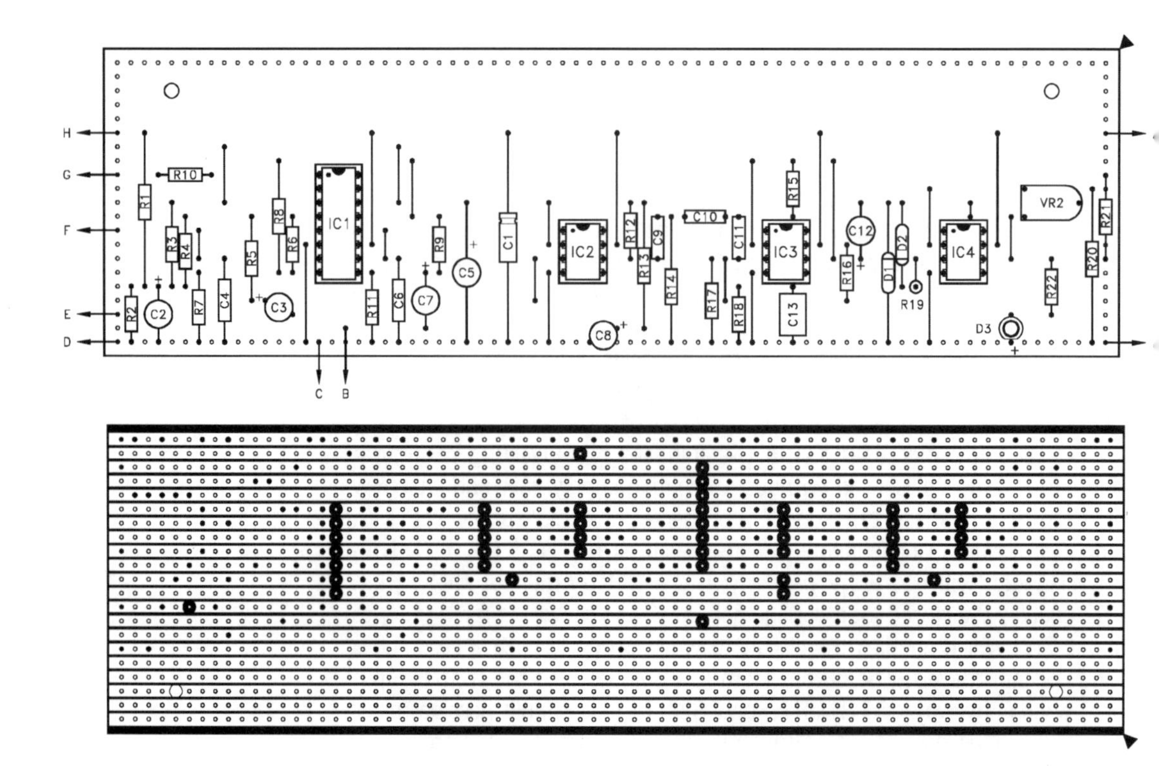

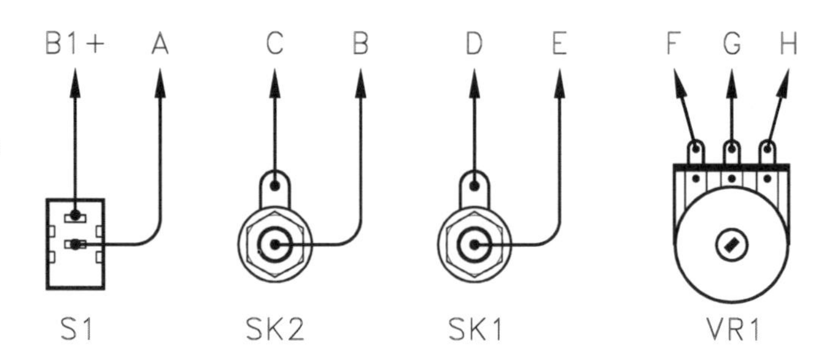

Top Figure 20.3 Component layout of the dynamic noise limiter

Middle Figure 20.4 Underside view of the board. It measures 74 holes by 21 copper strips

Below Figure 20.5 Hard wiring of the dynamic noise limiter

Construction of the board should is reasonably easy, although due to its relative complexity it is not a good choice for a complete beginner. The CA3140E used for IC4 has a MOS input stage, and it therefore requires the usual anti-static handling precautions. D1 and D2 are germanium diodes, and consequently require extra care to avoid damage due to overheating when they are being soldered into place. An electrolytic capacitor can be used for C13 (with the positive terminal connected to D2 and R19), but a non-electrolytic type is likely to give more reliable and predictable results. Note that C13 should have 10 millimetre (0.4 inch) lead spacing if it is to fit easily into this layout, whereas the other polyester capacitors have 7.5 millimetre (0.3 inch) lead spacing. It is best to use capacitors having long leadout wires for C4 and C6, such as Mylar or polystyrene components. D3 is shown as being connected on the board in Figure 20.3, but in the 'real thing' it will normally be mounted on the front panel of the case and hard wired to the circuit board.

In use

A dynamic noise limiter is used only during playback, and it simply connects between the signal source (VCR, camcorder or cassette deck) and the amplifier. With an input signal applied to the circuit, D3 should light up when the signal level is high. VR2 should be adjusted for the lowest gain (most counter-clockwise setting) that gives full brightness from D3 when the signal is close to its maximum level. It is possible that some signal sources will provide a slightly inadequate signal to drive the circuit properly. Increasing R12 to about 39k should cure this problem by boosting the gain in the side-chain.

The best setting for VR1 is a subjective matter, and it is just a matter of experimenting a little to find the setting that you like best. When the unit is processing music it is probably best to err on the side of caution and not try for very high degrees of noise reduction. A much higher level of noise reduction can be used when the unit is only processing speech signals.

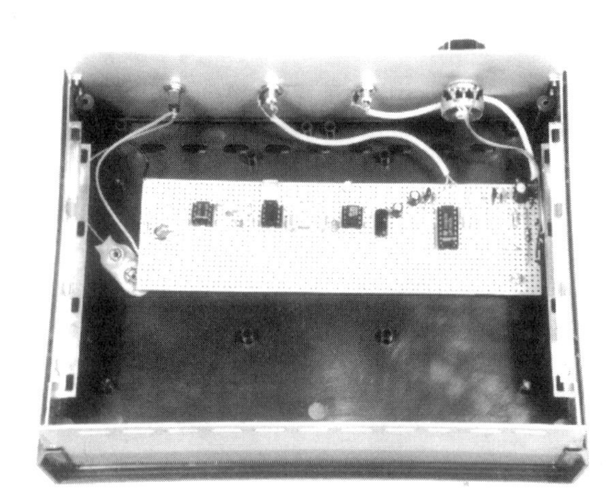

Photo 20.1 The interior of the dynamic noise limiter. Keep layouts simple and straight forward.

Photo 20.2 The finished dynamic noise limiter fitted in a plastic and metal case.

List of components

Resistors (all 0.25 watt 5% carbon film)

R1,2	3k9 (2 off)
R3,4,5,6,8,9	22k (6 off)
R7,11	4k7 (2 off)
R10,21	56k (2 off)
R12	15k
R13,15,20	3k3 (3 off)
R14	5k6
R16,17	100k (2 off)
R18	270k
R19	10k
R22	1k5

Potentiometers

VR1	100k lin carbon rotary
VR2	22k min hor preset

Capacitors

C1	100u 16V axial elect
C2	100u 10V radial elect
C3,8,12	1u 50V radial elect (3 off)
C4,6	4n7 mylar (2 off)
C5,7	10u 25V radial elect (2 off)
C9,10,11	22n polyester, 7.5mm lead spacing (3 off)
C13	1u polyester, 10mm lead spacing

Semiconductors

IC1	LM13700N or LM13600N
IC2,3	TL071CP (2 off)
IC4	CA3140E
D1,2	OA91 (2 off)
D3	Red panel l.e.d.

Miscellaneous

S1	S.P.S.T. min toggle switch
B1	9 volt (PP3 size)
SK1,2	Phono socket (2 off)

Metal case, 0.1 inch pitch stripboard having 74 holes by 21 copper strips, 8-pin d.i.l. holder (3 off), 16-pin d.i.l. holder, control knob, battery connector, wire, solder, etc.

Index

Introducing Digital Audio

2nd edition

Ian Sinclair

160 pp * 64 line drawings
ISBN 1870775 22 8

- ✱ For enthusiasts, technicians and students
- ✱ Techniques explained non-mathematically
- ✱ Covers CD and DAT
- ✱ Oversampling and bitstream methods
- ✱ Philips DCC and Sony Mini Disc
- ✱ Glossary of terms

Digital audio involves methods and circuits that are totally alien to the technician or keen amateur who has previously worked with audio circuits.

This book is intended to bridge the gap of understanding for the technician and enthusiast. The principles and methods are explained, but the mathematical background and theory are avoided other than to state the end product.

This second edition has been updated to include sections on oversampling methods and bitstream techniques. The opportunity has also been taken to include information on Philips DCC and Sony Mini Disc, and to add a glossary of technical terms.

Reviews of first edition
'Readable and informative' *Home & Studio Recording*
'Well worth a read ... the writing is clear and unambiguous' *The Gramophone*

PC Publishing

Tel 01732 770893 • Fax 01732 770268 • email pcp@cix.compulink.co.uk
Website http://www.pc-pubs.demon.co.uk